THE LMS WAGON

R. J. ESSERY
and
K. R. MORGAN

DAVID & CHARLES
NEWTON ABBOT LONDON
NORTH POMFRET (VT) VANCOUVER

ISBN 0 7153 7357 9
Library of Congress Catalog Card Number 76-54074

Set in 10 on 11 point Plantin
by Ronset Limited, Darwen

ISBN: 978-0-7153-7357-6

Published in the United States of America
by David & Charles Inc
North Pomfret Vermont 05053 USA

Published in Canada
by Douglas David & Charles Limited
1875 Welch Street North Vancouver BC

CONTENTS

To the late W. O. (Bill) Steel.
Who awakened our interest in wagons, and whose name would have, but for his untimely death, appeared as one of the authors of this volume.

INTRODUCTION

GOODS wagons, of necessity, had to be of many different types to handle the many and varied traffics offered. During the period under review, in the face of increasing competition from road transport, almost everything manufactured or grown was carried in part by the railways. The railways' answer to road competition, in the shape of containers, added still further to the variety, and formed an increasingly important part of the railways goods equipment.

Before the grouping of the railways into the 'big four' in 1923, each of the larger companies had its own teams of wagon designers, and as each company had its own ideas on what a particular type of wagon should be like, it is hardly surprising that the variety of designs taken over by the LMS was enormous.

Due largely to the influence of the Railway Clearing House specifications, however, vehicles constructed in the last years before grouping were very similar in regard to size and carrying capacity, while the running gear had reached a stage in development which hardly changed throughout the whole period under review. Indeed it was not until the implementation of the British Railways modernisation plan in the late 1950s, that much significant change occurred in this direction.

It must be remembered that in the years between the two world wars the railways had many secondary cross country lines, serving relatively small communities, plus innumerable branch lines, which in the main served even smaller towns and villages.

At each of these places there were, besides the passenger facilities, extensive goods yards. They could usually boast a goods shed, where goods could be loaded and unloaded under cover; many had cattle docks, while a coal yard and two or three sidings with crane facilities were almost inevitable.

It is probably true to say that during this period almost as much traffic collectively was generated from such sources as the larger centres, and although some of the pre-grouping companies had built large capacity wagons, they were found to be somewhat limited in their sphere of usefulness; the 12 ton wagon, which could be relied upon to run loaded most of the time therefore became the accepted standard for most of the wagon stock.

Obvious exceptions to this rule, were those vehicles classed as *special wagons*. Built to carry very long or heavy loads, they were affected only by the limitations of the loading gauge and other route restrictions, and some very large and heavy vehicles were built in this category.

These vehicles apart, the LMS wagons followed the basic pattern outlined above. The story is one of gradual improvement, rather than of striking changes in design. It is not possible to break down design into definite periods, as it is with the contemporary LMS coaching stock, for although three basic types of construction were employed, they were used concurrently. In this volume, therefore, we have traced the development of each type of vehicle, noting the types of construction in each group as they occurred.

This volume completes a project, which started as an attempt to describe the liveries used by the London, Midland & Scottish Railway for its rolling stock. After dealing with the locomotives on this basis, it was realised that by comparison there was little published information of any sort regarding the passenger and goods vehicles. It was decided, therefore, to widen the scope of the volumes dealing with these vehicles, to include as much of the available information as possible.

In this volume we have attempted to describe the goods vehicles built by or for the LMS and its successor, British Railways, to LMS designs. The basis of our research has been the LMS lot and diagram books, and although for reasons explained later there are numerous gaps in the information, it is we believe as accurate and complete as it is possible to be at this date. At the same time we must point out that any errors or omissions are entirely our own responsibility.

We wish to record our grateful thanks to all those who have assisted us in this project, in particular to T. J. Edgington formerly of the London Midland Region for his untiring efforts on our behalf. In addition, it is fitting that we also record with thanks the names of others who assisted in this work: T. W. Bourne, George Dow, G. Fox, Don Rowland, A. Whitehead and R. E. Wilson.

Frontispiece: Derby Works on 29 December 1928. In the foreground is a sleeper trolley which was built without an allocated diagram number. The background shows a line of good wagons, mostly MR types, newly repaired, including an MR goods brake van and a standard LMS refrigerator van, both in a dark body colour. No satisfactory explanation of this darker body colour can be traced and the matter is discussed further in Chapter 2.

CHAPTER 1

WAGON DESIGN AND CONSTRUCTION

A GOODS wagon, from the designer's point of view, consisted of three main parts: the body, the underframe and the running gear.

When a new design was called for, the first requirement was a set of working drawings. Of these drawings the most important was the general arrangement (GA). This was a large scale (usually 1½in to 1ft) drawing, which showed side and end elevations, a plan view and sections of special features.

The body of most goods vehicles was constructed largely of timber, its design being governed by the traffic for which it was intended; and ranged from the simple one plank open wagon, without doors or other complications through to such vehicles as cattle wagons and brake vans with their relatively complicated bodywork.

The GA drawing showed the body in great detail, all sizes of timber being given, while all strapping, corner plates etc. were also shown, together with details of bolt and rivet sizes. From this it will be seen that the body could be built by reference only to the GA drawing, although some of the metal details also had their own detail drawings for the benefit of the departments responsible for manufacturing these items.

Underframes were of two main types, timber or steel. With the timber type it was logical to dimension this on the GA drawing, as the preparation of the timber from which it was constructed was undertaken at the same time, and in the same department, as the body timbers. Steel underframes were built as completely separate units, sometimes in a different works, and here there were two varieties of GA drawing. The first was fully detailed as before, whereas in the second type a separate underframe GA was issued, the main GA merely showing the underframe in outline only.

The running gear, in which term were included everything below solebar level, plus drawgear and buffers, was constructed completely from standard parts, used on many different types of vehicle. Many of these items conformed to the Railway Clearing House (RCH) specifications, for it was with the running gear that the RCH influence was felt most.

These items appeared on the GA drawing in outline only, together with a reference to their exact specification. Detailed drawings of each item were prepared for the use of the departments concerned with their manufacture. For example, an axlebox consisted of several separate details; a GA drawing of the complete axlebox was prepared showing how the various parts fitted together. From this, detail drawings were prepared which gave the dimensions of each individual part, and showed where machining had to take place, while for the body of the axlebox, a casting drawing was prepared for the benefit of the pattern makers and foundry staff.

Such items were produced in their thousands. Once designed it was only necessary to ensure that the correct details for the particular vehicle were specified, thus saving a considerable amount of design time.

Diagrams and Lots

IN addition to the wagon designers and builders, many other departments of the railway required details of goods vehicles, for example, traffic officers responsible for providing both the right type and quantity of vehicles at the various traffic centres. To have provided these officials with GA drawings would have proved both costly and cumbersome, as they gave far more detail than was required for these purposes. To overcome this problem small scale diagrams were prepared which gave the main dimensions, tare weight and carrying capacity, and telegraphic code etc.

These diagrams were bound together in book form, and formed a handy reference for the departments concerned. The diagram pages were numbered, and on most of the constituent companies of the LMS this formed the only means of identification. On the Midland Railway each diagram had been given a reference number, and this practice, perhaps not surprisingly, was perpetuated by the LMS. Thus we get 'goods brake van diagram 1656' or abbreviated D1656. In general, diagram numbers were alloted in ascending order, so that brake vans to D1919 were built later than the previous example.

The working drawings were amended from time to time, to incorporate modifications or improvements. Such drawings were referred to as a 'raised issue', and in general wagons built to this later issue retained the same diagram number.

Similarly the building of some types of vehicle was spread over a number of years, and here a completely new drawing was sometimes issued; again unless the vehicles to the new drawing differed considerably the old diagram number was usually retained.

In the early years of the LMS period, separate diagram numbers were sometimes issued to distinguish fitted (that is fitted with automatic vacuum brakes) from otherwise identical unfitted vehicles. Latterly the practice was to issue one diagram to cover unfitted, piped and fitted vehicles of the same basic type. In direct contrast, some types appear on different diagrams where the variations are so small as to defy identification apart from the running numbers. Nevertheless the diagram system provides a very convenient basis upon which to work. It must be borne in mind that considerable differences existed between individual vehicles to the same diagram numbers, while repairs accentuated these variations still further.

As in the Locomotive Department, schemes were sometimes prepared for new wagon designs, which in a few cases went as far as having GA drawings and diagrams prepared and issued, although no wagons were built. In the main these have been ignored, but in one or two cases, however, the diagram and lot books do not agree with photographic evidence, and these instances have been noted as far as is possible.

It should also be noted that the diagram numbering system applied only to the ordinary wagon stock. Special wagons and containers had their own diagram books, and were referred to by the page number only. Another point of interest is that whereas ordinary wagon stock of the pre-grouping companies retained individual diagram books, in the case of the special wagons all vehicles, both pre- and post-grouping, were combined into one volume.

It is impossible to describe the many minor variations that existed between individual vehicles. The information contained in this volume basically refers to the vehicles as built; for example many of the older vehicles were fitted with vacuum brakes by British Railways in the early days of the modernisation scheme. Many more were scrapped before being fitted, and no attempt has been made to define those which were so modified.

There was nothing sacrosanct about a goods wagon, and once its useful commercial life was finished it was often used as the basis for a departmental vehicle which bore little resemblance to the original. Covered goods vans with windows, steps, and even stoves were not unknown, while open goods wagons fitted with a peak roof for locomotive sand purposes, or with bodies cut down to a single plank were also quite common. (See plate 4a).

Finally a word of warning may not be out of place to modellers, who should not place too much reliance upon diagrams. They were not always to scale, and if they are the only source of drawing available, should be supplemented by as many photographs as possible.

'Lots' were a method used to control expenditure, and were in effect an authorisation to build a specified number of vehicles of a particular type. Lot books were compiled listing the lots in numerical order; the lot book with which we are concerned covered passenger and non-passenger coaching stock, wagons, special wagons and in part containers. The LMS lot book commenced at lot 1 and ran consecutively, so that brake vans to D1659 lot 241 were built later than the vehicles to lot 37 of the same type.

The first two categories have been dealt with in a previous volume*, and this volume covers the remaining three types of vehicle.

Construction

WAGONS were of many different types and it is difficult to detail the particular features associated with each type in a general form. Constructional methods used can be broken down into broad groups, which describe accurately the main features to be found in each group leaving the more detailed descriptions for each type of vehicle to be covered in their particular chapters.

Historically, the oldest form of wagon construction was the all-wood type. These vehicles had the body and underframe constructed almost entirely of timber, the only metal parts, apart from the running gear, being the corner plates, strapping, door hinges etc. Although in pre-grouping days a great variety of vehicles were built using this type of construction, the LMS employed it mainly for open wagons, of which many thousands were built; most of the cattle wagons also had this type of construction. It had the advantages of relatively cheap first cost, while repairs to the underframe could easily be undertaken at the many wagon repair shops scattered throughout the country. A typical wooden underframe is shown in fig 1.

Although the wooden underframe was relatively simple to produce, in service it suffered from the rough treatment meted out to goods vehicles, and it was therefore gradually superseded by the steel underframe, which although initially more costly

* *The LMS Coach*, Ian Allan Ltd.

Fig 1 Typical wooden underframe: 1, headstock; 2, solebar; 3, diagonal; 4, longitudinal; 5, cross member; 6, curb rail; 7, floor planks.

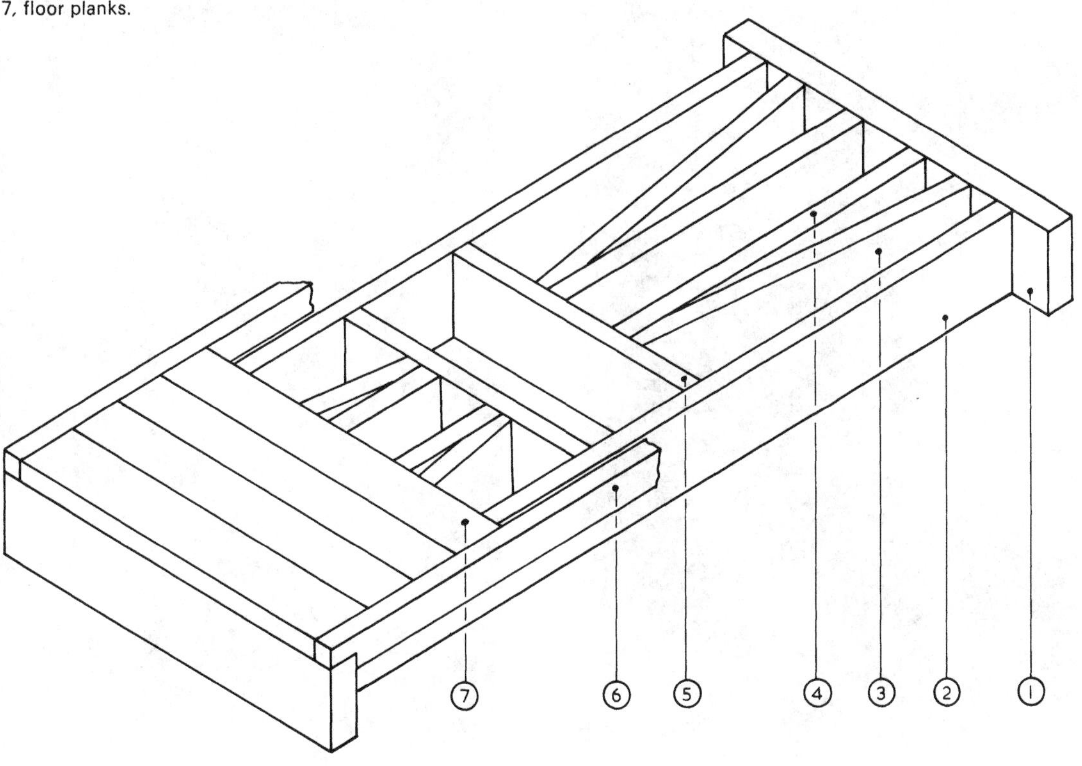

Below: Plate 1 Welded steel underframe for 12T wagon.

1276

PE10 70
1276

Top left: Plate 2A Three-link coupling on a 13T mineral wagon. V. R. Anderson

Top right; Plate 2B Double V Brake hanger on a wooden solebar mineral wagon. Note steel brake push rods and safety loop to prevent push rods dropping to the track in the event of a break in the linkage. V. R. Anderson

Bottom: These two photographs show the brake lever on a 13T mineral wagon in the 'brake-off' position, plate 2C and 'pinned down,' plate 2D. V. R. Anderson

was much more durable in service. Steel underframes were basically similar as regards the layout of the main members, but were formed from rolled steel channel sections. At first the main members were secured to one another by rivetted brackets, but were later of all-welded construction, which gave greater strength with a reduction in weight. Another important point was that whereas the wooden underframe was built up as part of the final assembly, in the case of the steel type it was built as a separate unit, and was delivered to the final assembly shop complete with all running gear. A steel underframe of the all-welded type is shown in plate 1.

Vehicles of both types of construction as outlined above were built by the LMS from its inception, the earliest major development being the vans with corrugated ends, the first examples of which appeared in 1924. These vehicles still had wooden sides, and the steel ends were lined with timber. Externally the type was very distinctive and the LMS later extended the principle to open wagons. With all these types of construction, the roof on covered vehicles was of timber construction covered with canvas.

Finally come all steel vehicles, and here there were three sub-types. The first was of conventional design, the timber planking being substituted by steel plates, secured to one another by rivetting. Many of the hopper wagons fall into this category, while some vans were also of this type of construction, although these were timber lined. The second type, of which the mineral wagons introduced in 1946 are representative, had an all-welded body, built in a jig and then secured to a conventional steel underframe.

Lastly come those vehicles, mainly of the bogie type, which fall into the special wagon category In general they had no separate body and underframe, but were built up as an integral girder formation, on which in many cases a timber decking was laid.

Containers, while not strictly vehicles, must also be mentioned, for constructional methods closely followed those of the wagon bodies; both wooden and steel versions were produced.

Dimensions and Details

CERTAIN features of the goods stock were standardised, and, to prevent repetition, details are given here. It can be taken that, unless otherwise stated, these features apply to all of the vehicles described in this volume.

Wheels

Wheels were either of the disc or spoked type, the latter being further divided into those wheels having solid spokes, and those where the spokes were of the split type. They tended to be used indiscriminately, especially in later years, and no attempt has been made to distinguish which type was fitted to any particular type of vehicle. It was common practice when repairing wagons to fit the first available wheels, and vehicles with one pair of disc and one pair of spoked wheels were an everyday sight. Unless otherwise stated wheels of 3ft 1½in or 3ft 2in diameter when new were fitted to all vehicles described in this volume.

Buffers and Drawgear

On unfitted stock the buffers extended 1ft 6in beyond the headstock, the couplings being of the three link type as illustrated in plate 2A. On piped and fitted stock the buffers extended 1ft 8½in beyond the headstock, the earlier vehicles using the same pattern buffers with a packing piece between the buffer stock and headstock. On the later vehicles special pattern buffers having the greater length were usually fitted. Couplings on the fitted stock were of the screw carriage type, while piped vehicles had either this type or in some cases the *instanter* pattern of three link coupling in which the centre link of T shape had long and short positions. An example of this type is shown in plate 19B. Buffers, in the majority of cases, were of the mechanically-sprung type, the earlier examples having a transverse leaf spring inside the underframe, the ends of which bore on the ends of the buffer shanks. These were superseded by the self-contained type, where each buffer had its own volute spring. Examples of some of the different patterns of buffer used are shown in plates 3B, 4A, and 4B.

Drawgear in all cases was of the continuous type. This, as its name implies, formed a continuous drawbar through each vehicle and hence throughout the length of the train. The main

These illustrations show close-ups for three different types of axleguard. Plate 3A, *top left,* shows a 20T goods brake van with the more sophisticated swing link compared with the wagon type axleguard in plate 3C *bottom left,* Plate 3B, *top right,* illustrates part of a vehicle from the special wagon book, 20T trolley. Note the heavy duty buffers, axlebox arrangement and spring and ratchet brake gear. Illustrated in Plate 3D, *bottom right,* is the brake gear arrangement on a 20T tube wagon. R. J. Essery, A. E. West (3B/D), K. R. Morgan

advantage of this system was that the underframe was relieved of most of the stresses set up by the pull of the locomotive. The way in which the draw gear for each vehicle was made up is clearly seen in plate 1.

Springs

The majority of goods vehicles were sprung by means of simple leaf springs. On unfitted stock these were of what is usually known as the Wagon Type, where the ends of the springs were retained in brackets bolted directly to the solebar, an example being shown in plate 3C. Brake vans, piped and fitted vehicles had a more sophisticated type, which had either swing links or small auxiliary springs, working on the ends of the main springs through J-shaped hangers, examples being shown in plates 3A and 12D.

Brakes

Before describing in detail the various types of brake gear employed on goods vehicles, an explanation of the terms used to describe the status of a wagon in this respect is necessary.

All goods vehicles had a hand brake, which in the case of the vehicles described in this volume was capable of being operated from either side, or in the case of brake vans from inside the vehicle. With the exception of the latter vehicles, brakes could only be applied when the vehicle was stationary or moving very slowly. Vehicles with a hand brake only were thus incapable of being braked under normal traffic conditions and were classed as *unfitted* by the railways. The use of these vehicles made running somewhat restrictive, and the presence at the top of steep gradients of the notice 'Stop to pin down brakes' is now a thing of the past. No more do drivers bring trains to a halt, and then draw forward down the gradient with the guard and brakesman (where provided) pinning down a predetermined number of brakes.

Some vehicles had, in addition to the hand brake as outlined above, a vacuum brake system. This generally operated on the same brake shoes as the hand brake; these vehicles when coupled up to the locomotive were thus directly under the control of the driver, and could safely be operated at higher speeds. Such vehicles were described as *fitted,* but during the LMS period, were in a minority. To supplement their numbers *piped* vehicles were built, which had a vacuum pipe

Plates 4A and 4B clearly show the two types of buffers fitted to wooden body 12T and 13T mineral wagons; 4B, *top right,* was used on the fixed end, 4A *top left,* on the end door end. Wagon label clip with label in position.
Plate 4C, *bottom left,* is a close-up of an LMS wagon number plate and 4D, *bottom right,* has the maker's plate on the same vehicle. This vehicle was photographed in 1971 and the L & S had been removed since nationalization.

These photographs illustrate just what can happen to a vehicle during its lifetime and indeed this low goods wagon began life as a 12T mineral wagon 45605 built at Derby in 1926. All V. R. Anderson

running the length of the vehicle with connections at both ends but without vacuum brakes. They could therefore be used to maintain continuity of the vacuum brake through several vehicles, though they were not in themselves under its control.

It will be clear that such vehicles were only of use where they were followed by further vacuum fitted vehicles, and to be effective had to be marshalled at the front end of the train. The number of completely fitted goods trains during the LMS era was quite small, most freights falling into the partially fitted category, where the fitted portion was followed by a tail of unfitted vehicles, the proportion of which decided the class of train. The use of piped vehicles helped to reduce the amount of shunting required when making up braked portions of trains; in the case of piped brake vans, a vacuum gauge and valve was provided so that the guard had control over the braking system.

Hand brakes on the majority of the ordinary goods wagons were of the side lever type, applied by depressing the long lever, and held in position by a pin passed through a hole in the lever guard above the lever, see plates 2C and 2D. On many of the earlier LMS vehicles two independent sets of brake gear were provided, one complete set on each side. This was known as double brake gear, and was an extension of pre-grouping practice, where in many cases brake gear was provided on one side only. This gear was necessary where bottom doors were fitted, as on mineral wagons, but as only one set of gear was in use at any one time it was very wasteful. To overcome this problem many ingenious systems were devised, the type usually known as the Morton being generally adopted by the LMS. With this type only one set of brake gear was provided, the side levers on each side being connected by means of a cross shaft beneath the vehicle, a clutch arrangement enabling either handle to be used to apply the brakes. The double brake can be distinguished by the two vee-hangers fitted on each side of the vehicle, one inside, the other outside the solebars. There were variations on the theme, especially in the case of longer vehicles where intermediate rodding was employed, but the basic principles remained the same. See plates 2B and 3D.

Vacuum fitted vehicles generally had a modified form of the Morton type, with two full sets of brake gear. The vacuum cylinder was arranged to operate through levers on the cross shaft in such a way that either brake (hand or vacuum) could be applied without affecting the other.

Brake vans, most of the special wagons, and some of the heavier types of ordinary wagon stock, were fitted with a screw down hand brake, operated from a hand wheel which was connected to a screw on which ran a nut or die block, in turn connected to the brake linkage. This had the advantage of applying greater pressure, and in the case of brake vans could be applied while the train was in motion.

Underframes

The solebars and headstocks of wooden underframes were usually 12in deep by 5in thick. Steel underframes had these members made out of channel section, the normal sizes of which were 9in deep by 3⅛in wide. Distance between solebars was 6ft 3in in both cases, width over headstocks being about 7ft 8in. The vast majority of the ordinary wagon stock was built to a common length of 17ft 6in, and where all of the above dimensions apply is referred to in the text as having standard underframes.

Ventilators

Ventilation on the majority of covered vehicles was by metal hood type ventilators, mounted centrally at the top of each end. In addition most vehicles had four torpedo type ventilators mounted on the roof, two on each side of the centre line. This arrangement is referred to in the text as standard.

Number and Builders Plates

All vehicles had cast iron numberplates, generally secured to the solebars towards the left hand end, above or near to the wheels. A standard LMS example is shown in plate 4C. LMS built vehicles had a small plate with date and works name thereon, on one side only, see fig 2. Wagons built by outside contractors usually carried their own makers plates, also on the solebars and often on both sides. A standard LMS example of these plates is shown in plate 4D.

Label Clips

These were fitted on both sides of the vehicle, either on the solebar or body, and were used to hold the labels describing the contents and destination of the wagons load; an example is shown in plate 3C.

Wagon Builders

The vehicles described in this volume were built both in the companies' own workshops and by outside contractors. Many of the early lots built by contractors, were indicated in the lot book by the word *trade*, no specific company being mentioned; later lots usually gave the name of the building company.

The LMS workshops concerned were: Bromsgrove—ex Midland Railway (containers only); Derby—ex Midland Railway; Earlestown—ex LNWR; Newton Heath—ex LYR; St Rollox—ex Caledonian Railway; Wolverton—ex LNWR.

Fig 2 LMS building plate

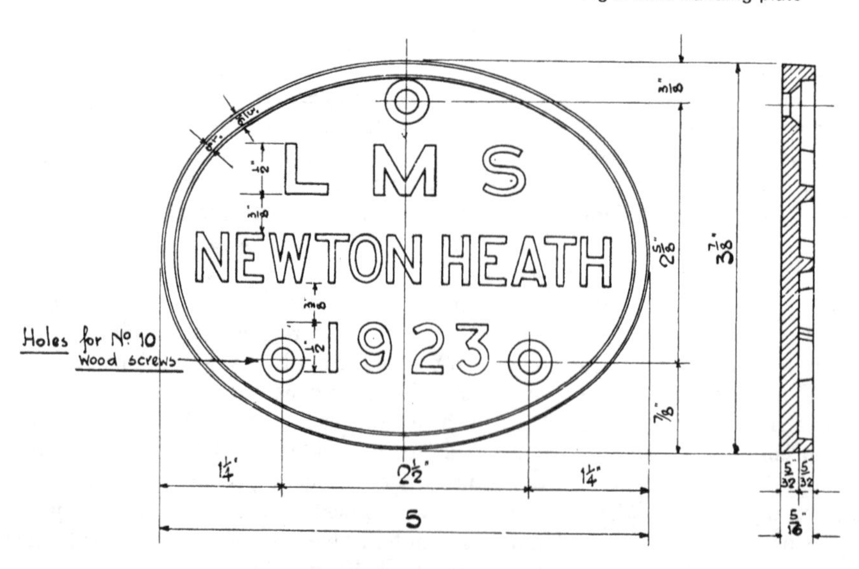

CHAPTER 2

LIVERY AND NUMBERING

WHEN one considers that the LMS adopted what was, as far as locomotives and carriages were concerned, the old Midland railway livery, it would have been surprising if the company had used anything other than the basic Midland livery for its wagon stock. This much is certain, but what is less certain is exactly what the Midland was doing in 1923. In this respect and at this juncture your authors freely admit that as far as livery is concerned there remain many unanswered questions, and this, coupled with the factor of weathering, different works interpretation of instructions and perhaps less emphasis on the importance of a corporate image for wagons, has tended to make the completion of this chapter rather more difficult.

However, with this admission and with the knowledge that it is probably unlikely that any further evidence will ever be forthcoming, the basic facts seem to be that the LMS used two colours, grey prior to 1936 and bauxite thereafter. The main problem is that more than one shade of grey appears to have been used. Midland practice in 1921 seems to have been to use a light shade for new construction and a dark grey, called smudge, for repaired wagons. The exact ingredients of either colour are not now known, and indeed it is unlikely that the latter colour remained consistent since it was a mixture of black and various paint scrapings. It was a standard practice for all paint work shops and decorating firms to use left over quantities of paint by tipping the contents into a paint drum and stirring in. The result was a mixture varying from day to day in colour which was often used for rough coating or undercoat work and the thicker consistency helped to make the mixture both a filler and undercoat.

The paint specifications in this chapter are those which appear in the 1935 edition of the official LMS schedule, and since the only other schedule we have seen is dated 1929 it would appear that up to this date Midland methods were followed completely. The grey specified in the 1929 schedule would have been darker than that shown in this chapter as Mix 5 since it called for 8lb of black in oil as against 3-4lb, while the ultramarine was omitted. Photographic evidence confuses the issue still further since some photographs taken before and during 1928 clearly show some vehicles, brake vans and refrigerator vans in particular, in a much darker colour.

Study of the frontispiece shows that the line of wagons clearly ex works have the cattle wagons in a much lighter shade than the refrigerator van; since the latter vehicle is brand new, the others being repaints, it would appear that the Midland policy was in this instance completely reversed. Your authors freely admit that they have not been able to throw any further light upon the subject and have been unable to find an example of this dark grey in anything other than a vehicle which has been part of the background view.

The official change to bauxite in May 1936 was preceeded slightly by a change to what was to become the more standardised type of livery and some new construction, outshopped in the grey body colour, was lettered in the new style before bauxite became general for all bodywork. Even so some vehicles were re-lettered but not repainted, so that until after 1948 it was possible to see vehicles in a weathered grey body colour carrying the post-1936 lettering style. Some of these vehicles even had traces of the pre-May 1936 lettering still visible.

During the War a further livery change occurred, many wooden-bodied vehicles being built without the benefit of paintwork on the woodwork except where lettering was placed, and here painted patches were used with the livery detail of numbering and letters painted on to the bauxite painted patches. See plate 12B.

Below solebars everything was painted black but the solebar itself was painted the relevant body colour. Above the bottom of the solebar there were exceptions to this rule. Buffer heads and shanks, coupling hooks, links and brake lever guards were painted black; a patch giving oiling date and district numbers was painted slate grey.

Roof Colour

The colour of van roofs is quite a problem and readers are referred to the available written evidence. The LMS carriage and wagon department, schedule of paints and formulae, states:—

> In the case of wooden roofs the same procedure must be followed as in the case of carriage roofs.

Before 1935 the specification differed in that the final operation was to apply four coats of a mixture of equal parts of protective white paint and tarpaulin dressing and although the 1936 specification shows an aluminium colour type paint was used, we feel this is rather misleading since after a short time in service, weathering conditions would rapidly turn the ex-works roof colour muddy grey, darkening as the time lengthened from leaving the works.

Service Vehicles

While the foregoing applied to revenue earning vehicles, there was also a considerable amount of service stock. The LMS did not add to this to any great extent for several years, and LMS livery policy for this stock needs clarification. Indeed researchers could be forgiven if they believed no such policy existed. Nowhere within the available written evidence is there any mention of a livery for service stock, and photographic evidence suggests the existence of at least three other colour schemes. Because the number of vehicles was fairly small and their duties were of a humble nature, they seem to have almost escaped notice. This summary, while not conclusive, is believed to be fairly accurate.

PLOUGH BRAKE VANS—Probably slate grey body.

BALLAST WAGONS—Wartime construction unpainted but earlier construction had red oxide body work.

SLEEPER WAGONS—As ballast wagons.

LOCO COAL WAGONS—Normal freight stock livery.

BOGIE HOPPER COAL WAGONS—Normal freight stock livery.

CHAIRED SLEEPER TROLLIES—Normal freight stock livery.

CRANE MATCH WAGONS—Some black all over, others slate grey body and black underframes.

Painting Methods

ALTHOUGH the painting of goods stock did not receive quite so much attention as contemporary coaching stock, nevertheless the 1935 painting schedule is still comprehensive and shows that the painting methods, at least for new stock, was not as haphazard as is often supposed. The main points of divergance between the two were that fewer operations were involved in the case of goods stock, and no set time was specified between the operations.

Wooden Bodied Wagons

Before assembly started, all of the previously prepared timber received one coat of primer (mix 1) on all edges which were inaccessible after assembly. This was allowed to dry hard before any further work was done. Similarly all steel details had rust, scale etc removed by chipping, scraping and wire brushing. Oil and grease were then removed with white spirit and one coat of bauxite (mix 2) applied immediately. Where welding was employed, the weld was first freed from all slag, and the adjacent areas were wire brushed; after examination an area extending for two inches on each side of the weld was given one coat of primer (mix 3A) followed by one coat of bauxite (mix 2) all over.

After assembly, all of the exterior woodwork was given one coat of knotting to all knots. This was followed by one coat of primer (mix 1 or 4A) and any imperfections in the woodwork were stopped up with putty. All of the exterior ironwork was touched up where necessary with mix 2, and this was followed by a further coat of the same mixture or one coat of mix 2A. Finally the exterior of the body, solebars, headstocks and buffers was given two coats of grey (mix 5), or one coat if flow painted. After 1936 this operation was carried out using bauxite (mix 2B). While the specification does not detail flow painting, this term was used by London Transport to describe a method used to paint buses. The paint was poured into a distributor which allowed it to flow out on to the top edge of the bus. The paint then flowed down the bus sides into recovery drains.

The undergear and inside faces of the frame received one coat of bauxite (mix 2A), followed on the undergear by one coat of finishing black enamel.

The exterior of the roof on covered vehicles first had one coat or primer (mix 1 or 4A) applied; it was then stopped up with putty and one coat of jointing paste applied. The canvas was next stretched on and bedded down. One coat of a mixture consisting of equal parts of boiled linseed oil and jointing paste was then applied, followed by three coats of roof paint (mix 6).

It will be noticed that nothing is said in the foregoing specification regarding the interiors of vehicles, and in fact the only instance where such information is quoted is in respect of the brake vans, which after having the treatment already described had a coat of knotting applied to all interior knots. The ceiling then had two coats of white undercoat (mix 7), followed by one coat of finishing white (mix 7A). The sides and ends were given one coat of primer (mix 1) followed by one coat of green undercoat (mix 8) and one finishing coat of green (mix 8A). The lettering was then applied, in all cases using white paint (mix 9). The specification states that this was to be done by spraying, though as plate 18A shows, much of this work was in fact hand applied.

It will be noted that the specification does not mention the painting of wooden underframes, probably because when it was issued new vehicles of this type were fairly rare. It does state that repaired vehicles of this type were to have one coat of acid resisting black enamel on the top faces of all main members.

On the matter of repaints, the specification is equally vague, merely stating that the condition of the wagon and the facilities available must be taken into account when determining methods to be used, which perhaps serves to show that the subject did not receive quite so much attention as did passenger stock.

Steel-Bodied Wagons

The procedure adopted here was very similar to that already described for the steel details of the wooden-bodied wagons. The only point to be noted is that the interior of hopper wagons was finished in one of two ways. The first specified that one coat of bauxite (mix 2) and one of bauxite (mix 2A) was to be followed by one coat of finishing black enamel. The second method was to apply one coat of bauxite (mix 2) followed by two coats of bituminous paint (mix 10). An exception to this rule applied to grain hoppers, which were to be left unpainted.

Containers

Although the painting methods for containers were given in a completely separate part of the specification, it is hardly surprising that much of the information duplicated that already given for wagon stock. Unless otherwise stated, the information already given applies equally to this section.

Open containers were painted grey, or after 1936 bauxite, the only point to be noted being that the steel type was given one coat of steel primer (mix 3) inside and out, followed by two coats of grey (mix 5) on the exterior and one coat on the interior surfaces. The underside of the floor was given one coat of primer, followed by one coat of acid resisting black enamel, the wooden type receiving the enamel only, apparently without the benefit of a priming coat.

The specification for covered containers was divided into four groups, viz, wood or steel construction, painted white or lake. Taking the white painted wooden type first, the specification stated that after the preliminary operations one coat of white undercoat (mix 7) was to be applied, followed after stopping up by two further undercoats. Two coats of glossy white enamel followed, after which the lettering was applied using black air drying enamel. The lettering was then varnished using carriage stock exterior finishing varnish. The underside of the floor was given one coat of acid resisting black enamel, while the interior received two coats of knotting.

The lake liveried wooden containers received one coat of primer (mix 4A), stopping up being done with hard stopping (mix 11). One coat of lead colour undercoat (mix 4) was followed by one coat of lake undercoat (mix 12) and one of standard lake (mix 13). The specification stated that lettering was then to be carried out using yellow paint (mix 14), (but see livery details regarding this point). Finally the exterior was varnished with two coats of carriage stock exterior finishing varnish, flatted down lightly between coats with pumice dust. The underside was dealt with as described above, while the roof in both cases was dealt with exactly as already described for the wagon stock.

The covered steel containers had a somewhat different treatment, the first part of which can be considered for both liveries. All of the steel panels were covered with panel wash (mix 15), which was left for 30 minutes and the panels were then washed in warm water and thoroughly dried. This process had to be carried out at least one day before proceeding further. One coat of steel primer (mix 3) was then applied to both surfaces. The white containers were painted exactly like the wooden variety; the lake containers, however, only had one coat of lake undercoat (mix 12) followed by one coat of lake (mix 13), lettering and varnishing being as for the wooden variety.

The interior, where it was unlined, was given two coats of white synthetic enamel; where a wooden lining was present the interior of the shell was given one coat of bauxite (mix 2), the plywood panelling being given two coats of knotting.

Where the roof was of galvanised material it was given a coating of zinc wash (mix 16); this was allowed to dry and wiped clean. Where mild steel was used it was treated as already described for the side panels, after which in both cases two coats of steel primer (mix 3) were applied to the outside and one to the inside. The exterior of the roof was then given two coats of roof paint (mix 6). The underside of the floor received one coat of steel primer (mix 3) and one coat of acid resisting black enamel.

Paint Specifications

MIX 1 SMUDGE PRIMER

No quantities were specified for this item. It consisted of smudge and zinc white, composite pigment, in oil and suitably tinted.

MIX 2 BAUXITE PAINT (UNDERCOAT)

Boiled linseed oil	8lb
White spirit	6-10lb
Liquid drier	2–4lb
Bauxite residue in oil	82lb

MIX 2A BAUXITE PAINT (SECOND COAT)

Mixture number 2	100lb
Black in oil	6lb

MIX 2B BAUXITE PAINT (FINISHING COAT)

This is not shown in the specification, but its probable composition is given below:

Mixture number 2A	90lb
Mixing varnish	10lb

MIX 3 PRIMER FOR STEEL

Oxide of iron, in oil, type R (red shade)	88lb
Zinc oxide white in oil	2lb
Aluminium powder (fine varnish powder)	10lb
Raw linseed oil	10lb
Mixing varnish	26lb
Genuine turpentine	16–20lb
Liquid drier not more than	4lb

MIX 3A PRIMER FOR WELDED JOINTS

Mixture number 3	80lb
Aluminium powder (fine varnish powder)	10lb
Mixing varnish	10lb

MIX 4 LEAD COLOUR UNDERCOAT

Protective white paint paste	112lb
Liquid drier	9–12lb
White spirit	26–30lb
Black in oil	9–10lb
Raw linseed oil	4lb

MIX 4A WOOD PRIMER

Mixture number 4	80lb
Aluminium powder (fine varnish powder)	10lb
Mixing varnish	10lb

MIX 5 LEAD COLOUR FOR WAGONS

Zinc white, composite pigment in oil	112lb
Boiled linseed oil	60lb
Black in oil	3–4lb
Liquid drier	2–5lb
Mixing varnish	9lb
Ultramarine blue in oil	3–4lb

MIX 6 ROOF PAINT

Protective white paint paste	56lb
Thickened linseed oil	4lb
Boiled linseed oil	7lb
Mixing varnish	7lb
White spirit	8–12lb
Black in oil	8lb
Aluminium powder (fine varnish powder)	7lb
Liquid drier not more than	4lb

MIX 7 WHITE UNDERCOAT PAINT

Zinc white, composite pigment in oil	112lb
White spirit	27lb
Paste driers in oil	3lb
Mixing varnish	4–7lb

MIX 7A WHITE PAINT FOR INTERIORS

Mixture number 7	90lb
Mixing varnish	10lb

MIX 8 GREEN UNDERCOAT FOR GOODS BRAKES

Green, middle brunswick, in oil	36lb
Raw linseed oil	10lb
Liquid drier	2lb
Boiled linseed oil	12–16lb
Mixing varnish	4lb

MIX 8A FINISHING GREEN FOR GOODS BRAKES

Mixture number 8	90lb
Mixing varnish	10lb

MIX 9 WHITE PAINT FOR LETTERING

Zinc white oxide in oil	60lb
Titanium white in oil	150lb
Zinc white, composite pigment in oil	230lb
Raw linseed oil	10–20lb
Paste driers in oil	40lb
Gold size, type B (light)	10–30lb
Mixing varnish	40–60lb
White spirit	40–50lb

MIX 10 BITUMINOUS PAINT

Blown bitumin	50lb
White spirit	50lb

MIX 11 HARD STOPPING

Enamel filling	112lb
Gold size, type A (dark)	4 parts
Genuine turpentine	1 part

(sufficient to bring to suitable consistency)

MIX 12 UNDERCOAT FOR LAKE

Oxide in iron, in oil, type R (red shade)	100lb
Liquid drier	4–6lb
Mixing varnish	28–30lb
Genuine turpentine	12–14lb

The above mixture produced a brown undercoat, this was taken and mixed as below to produce the lake undercoat.

Mixture as above	95lb
Black in oil	5lb

MIX 13 STANDARD LAKE

Standard LMS lake (paste form)	12lb
Mixing varnish	4lb
Genuine turpentine	3–5lb
Liquid drier	1–3lb

MIX 14 YELLOW PAINT FOR LETTERING LAKE CONTAINERS

Zinc white, composite pigment, in oil	2lb
Lemon chrome in oil	12oz
Liquid drier	2–3oz

MIX 15 PANEL WASH FOR STEEL PANELS

Phosphoric acid	2 gallons
Methylated spirit	8 gallons

MIX 16 ZINC WASH FOR GALVANISED PANELS

Methylated spirits	6 gallons
Toluol	3 gallons
Spirits of salts (hydrochloric acid)	½ gallon
Carbon tetrachloride	½ gallon

Numbering

The pre-grouping companies in general had no real system of wagon numbering. In most cases they merely started at one, and numbered their stock consecutively, without any regard for type, while new vehicles often took the numbers vacated by withdrawals. There were exceptions; both the LNWR and MR, for example, appear to have allotted the lowest numbers in their wagon fleets to the brake vans. At the time of the grouping the LMS took over about 305,000 wagons from its constituent companies, and renumbered them, in most cases by a straight addition to their old numbers, thus perpetuating the random numbering system. A point of interest here is that some vehicles were renumbered without being repainted, these vehicles often retaining their pre-grouping livery well into the 1930s.

At first, new construction continued to take vacant numbers within the range thus laid down, some vehicles being allocated in small blocks of numbers, while some of the early brake vans took block numbers higher than any previously allocated. In 1934 the block numbering system was adopted for all new construction, the lowest number under this scheme being 400000. Unlike coaching stock, which was completely renumbered at this time, existing wagons were not affected. The probable reason for this was that to trace and replace the cast iron numberplates of all the wagons was considered to be too mammoth a task even for the tidy minded LMS Railway.

Unfortunately a large number of the wagon records were destroyed by a fire at Derby works some years ago, and it is not possible to give complete numbering details in this volume. Even if it were possible, it is felt that few readers would be interested in looking at a list of over 54,000 numbers for one type of open wagon. Here we give a selection of the numbers carried by these vehicles, the block numbers being quoted in full as far as is possible.

Outline of LMS Wagon Numbering Scheme 1923

Midland Railway

Retained original numbers within the block 1–129000*

*This statement requires qualification, inasmuch as Midland brake vans carried an M prefix to their numbers, this practice being dropped after grouping.

Lancashire & Yorkshire Railway

Renumbered by the addition of 130000 to their original numbers, between 130001–169999

Glasgow & South Western Railway

Renumbered by the addition of 170000 to their original numbers, between 170001–191999

London & North Western Railway

Renumbered by the addition of 200000 to their original numbers, between 200001–278999 with the following exceptions:

279000 added to the numbers of service vehicles up to 279999

280000 added to the numbers of brake vans up to 281999

282000 added to the numbers of service vehicles up to 284999

Caledonian Railway
Renumbered by the addition of 300000 to their original numbers between 300001–352999, with the exception of brake vans and service vehicles, which were renumbered by the addition of 353000 to their original numbers up to about 356999.

The smaller companies' stocks were renumbered into blocks, in this case no account being taken of their former numbers. The vehicles were renumbered as they became available, taking the lowest number vacant within the block, the scheme for these vehicles being:

North Staffordshire Railway
Numbered between 192000–199999

Furness Railway
Numbered between 285000–291999**
**This number block presumably also included Maryport & Carlisle stock.

Highland Railway
Numbered between 292000–299999

After nationalisation LMS vehicles retained their numbers, prefixed with the letter M, service vehicles being prefixed DM. Some vehicles of LMS design built after 1948 were given BR lot numbers, and appeared in the BR diagram book. These vehicles were given numbers in the BR series with a B prefix.

Insignia Details

UNLIKE passenger stock, where most insignia were applied by transfers, those on the goods vehicles were painted on, in most cases by hand. It is hardly surprising that considerable variations existed, both in the shape and positioning of the characters.

This was particularly true during the earlier part of the period under review, for with the change to bauxite livery a standard position for each item of insignia was laid down. The following paragraphs attempt to describe the insignia used, and the standard positions used after 1936, supplementing this with additional notes in each chapter for the various types of vehicle involved.

Company Initials
During the period when grey livery was in use company initials were of several different sizes. In general each type of vehicle employed a particular size of letter, the largest of which was 18in high. The shape and sizes most used are shown in fig 3, but considerable variation of shape occurred, particularly in respect of the S.

With the change to bauxite livery 4in letters became the rule, the standard position being at the left hand end of the body, above the carrying capacity and running number. As a wartime measure the height of the letters was reduced still further to 3in.

Vehicle Number
The vehicle number was carried on the cast plates as described in Chapter One, and in addition was painted on the body. On the majority of vehicles they were 4in high, but on the brake vans at first they were 5½in high, and were painted on a black panel having a white surround, see fig 3. This latter practice was of MR origin, and it is worth noting here that it was not universally carried out on the brake vans of pre-grouping origin; LNWR vehicles in particular seem to have had plain numbering from the start. The 1936 standard position was at the left hand end of the body, below the company initials and carrying capacity. Wartime measures again stated that they were to be reduced to 3in in height.

Carrying Capacity
On the majority of vehicles this appears, at first, to have been confined to the cast number plates. There were naturally exceptions, and in these cases the usual marking appears to have consisted of the capacity in figures, followed by the word Tons in full. With the introduction of the bauxite livery, the capacity in figures was followed by the letter T, both being 3in high. It was again stated that this was to be reduced to 2in as a wartime measure, the standard position being at the left hand end, between the company initials and vehicle number.

Tare Weight
On brake vans, this was at first given as the weight in figures, followed by the word Tons in full. This was later abbreviated to the letter T only. On the other vehicles the weight was given in figures only, as tons cwts and quarters with a small hyphen between each figure. The 1936 specification stated that the weight was to be given as tons and cwts only, the standard position being at the right hand bottom corner of the body, or on the solebar for vehicles without a body. Height of figures was 3in reduced to 2in as a wartime measure.

Non-Common-User
This was confined to vacuum fitted stock and

LMS

(a)

(b)

Fig 3 Livery insignia details: (a) Company Initials. Various sizes in the early period, but standardised at 4in high in 1935/6. Drawing shows typical shape of characters though some variation is observable in photographs.
(b) Brake Van Number Panel. Used on all standard and some pre-grouping vehicles 1923–35.
(c) Vacuum Release Cord Star. As used initially, star with blunt points adopted later.
(d) Metro Gauge Mark.
(e) Vacuum Fitted Vehicle (X). As used up to about 1935.

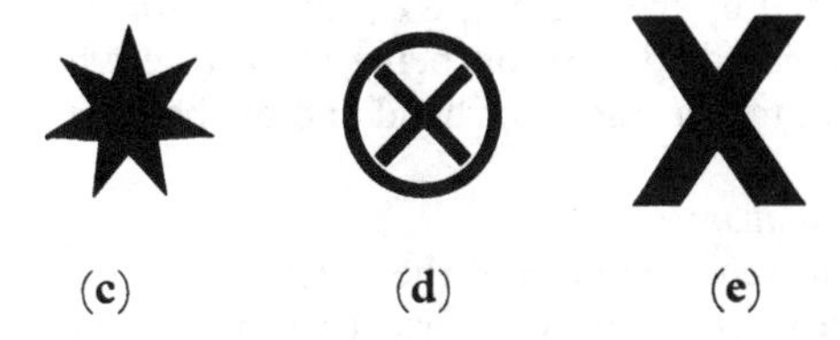

(c) (d) (e)

EXAMPLES OF MARKINGS USED ON SPECIAL TRAFFIC WAGONS

These varied considerably, especially where pre-grouping vehicles were concerned, spacings etc being modified to suit the configuration of the particular vehicle. The examples shown are all used on standard LMS designed wagons.

LARGE SINGLE

BULK GRAIN

LOCO

STEAM LONG LOW

BANANA TUBE

MEAT

BEER VAN

DOUBLE

VENTILATED VAN

PLATE

REFRIGERATOR

INSULATED

special wagons and took the form of a letter N 4in in height (3in in wartime), the standard position being at the lower edge of the body at each end.

Vacuum Fitted Stock

The first marking for these vehicles took the form of a large letter X generally 12in in height. The date of introduction of this symbol is uncertain, but appears to have been in the late 1920s; it should not be confused with the same symbol which was used by some pre-grouping companies to denote common user vehicles. This marking was supplemented by a seven pointed star, placed by the vacuum release cord. The initial shape of this star was as shown in fig 3, being modified later to have much blunter points. The X was later replaced by the marking XP; this was 4in in height, the standard position being above the wheelbase markings at the right hand end of the body.

Wheelbase

This took the form of the letters WB followed by the wheel-base in figures for feet and inches. It appears to have been little used before 1936 and thereafter was used mainly on fitted stock. Height of the symbols was 2in, the standard position being at the right hand end of the body, under the XP marking.

Brake Overhaul, Lifting and Painting Dates

These were stencilled on the solebars in 1in high figures, being prefixed BO, L and P respectively. These were followed by the date in figures, ie 11–10–34 and suffixed by a number which indicated the works involved. The brake overhaul date was applied near to the vacuum cord star, the other two items being at the left hand end.

Oiling Date

This took the form of a black painted panel 10in × 5½in, positioned on the solebar above the right hand wheel. Painted on it in white letters were the words Date Oiled and District No., the relevant information then being stencilled on.

Wagon Code

This was painted at the left hand end of the solebar, height of the figures being 2in. It does not appear to have been very much used before 1936, although some vehicles had metal plates with the code on before this date.

Metro Gauge Mark

This again appears to have been little used on the wagon stock; it is illustrated in fig 3, and indicated that the vehicle concerned could work over the so-called Widened Lines of the London Transport Metropolitan route.

In addition to the foregoing, various descriptive words such as Banana, Gunpowder etc were used, while there were also some special symbols on such vehicles as mineral wagons. Details of these markings will be found in the chapters which cover the vehicles concerned.

CHAPTER 3

GOODS BRAKE VANS

GOODS brake vans have been chosen to occupy the leading position in this volume for two reasons. First it was the position they occupied in the wagon diagram book, and second, whatever other vehicles may have gone to make up a goods train, in LMS days inevitably at the tail end there came a brake van. These vehicles performed a vital function in assisting the enginemen to control the train, for as will become apparent in subsequent chapters, the unfitted vehicles far out-numbered those fitted with the automatic vacuum brake. Under these conditions it was vital for the guard to know the road, to anticipate the driver's needs, and to prevent the train from taking control on long downhill gradients, with possibly disastrous results.

The vans also displayed lamps which informed the drivers of trains on adjacent lines that the train they were overtaking was a goods, on which line it was travelling, and which also served to show the signalmen and the engine crew at the front that the train was complete. The positioning of these lamps is of interest. On single and double lines of track, a tail lamp was carried on the lower part of the verandah end, slightly offset from the centre line of the vehicle. In addition two side lamps were carried, one on each side of the body, higher up, either on the bodyside or on the verandah corner posts. These lamps normally showed a red aspect to the rear and the side lamps white towards the front.

Where there were three or four running lines,

the same arrangement was used if the train was on the fast line, but if the train was on the slow, goods, or loop line, the side lamp on the side of the van nearest to the fast line displayed a white light to the rear. On goods or loop lines adjoining four main lines the side lamps were removed, leaving only the tail lamp.

Where a goods train was being assisted at the rear end by a banking engine, the tail lamp was removed leaving the side lamps in position. If the banking engine was pulling a brake van, the side lamps of the train brake van were also removed, and side lamps were placed on the van being pulled by the banker. When the banking engine left the train at a signal box, the tail lamp and side lamps (where applicable) were replaced within view of the signalman. If the banking engine left the train between signalboxes, the side lamps were replaced immediately, but the tail lamp was not replaced until the next signalbox had been passed, this procedure being adopted to bring to the attention of the signalman the fact that the banking engine was still in the section.

The foregoing procedure may have been carried out several times in the course of a single journey, and although it may seem lengthy and complicated, it serves to emphasise how necessary it was for a goods guard to know the road intimately.

Brake vans, more than any other vehicles, characterised their owners, and a study of the pre-grouping scene shows the tremendous variety that existed among them. One common feature that emerges, however, is that among the larger companies the 20-ton vehicle had become the standard, and without exception this was the weight of all the vehicles built for normal duties by the LMS.

As in other departments, Midland Railway practice predominated, the first LMS design being practically identical to the final MR type, although allocated a new diagram number. The main feature of the design was a centrally placed body with identical verandahs at each end, the roof extending over the verandahs to the full length of the underframe. Although later vehicles were longer, the basic outline remained virtually unchanged to the end of the LMS era.

The main dimensions remained remarkably consistent: body length was 13ft 4in inside (13ft 5in on later vehicles), width over body was 7ft 6in and over grab handles 7ft 10½in. Height to the centre line of roof was 11ft 2in. All had steel underframes, the main members of which were 12in deep.

The first LMS vehicles to D1659 were unfitted, and were virtually identical to the final Midland Railway vans, some built in 1923 being counted among that company's stock. The body featured a wooden floor, on which was built a

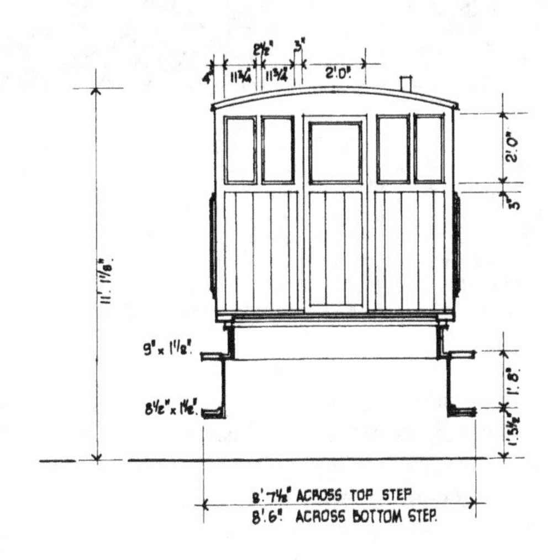

Fig 4 D1659 Goods brake van

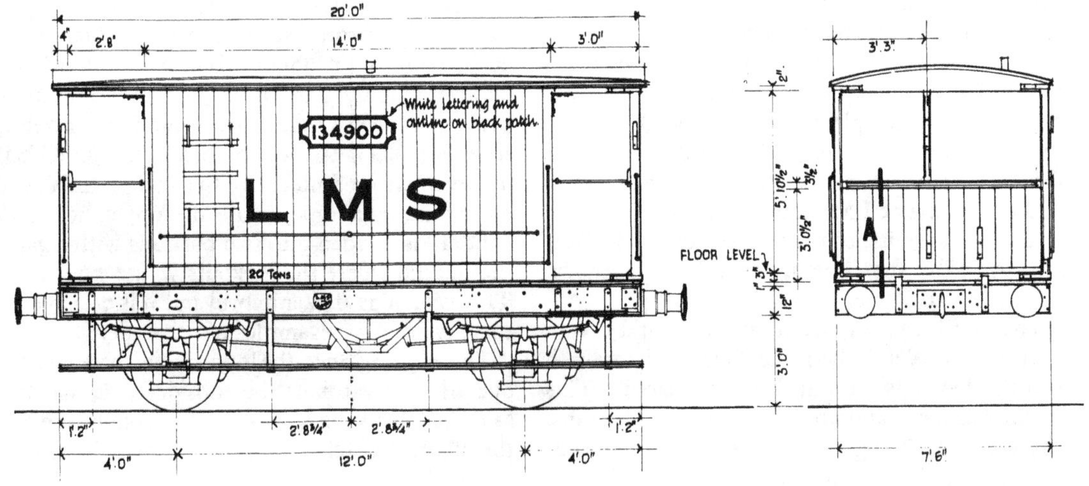

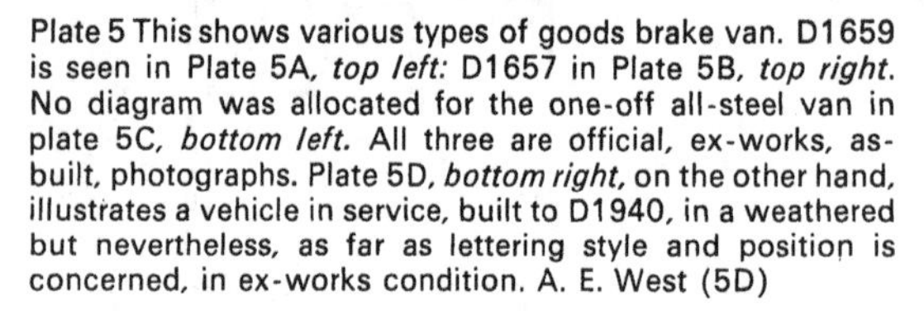
Plate 5 This shows various types of goods brake van. D1659 is seen in Plate 5A, *top left:* D1657 in Plate 5B, *top right.* No diagram was allocated for the one-off all-steel van in plate 5C, *bottom left.* All three are official, ex-works, as-built, photographs. Plate 5D, *bottom right,* on the other hand, illustrates a vehicle in service, built to D1940, in a weathered but nevertheless, as far as lettering style and position is concerned, in ex-works condition. A. E. West (5D)

heavy wooden framing; this was then planked on the outside to present a flush finish, the planks being vertical. They were 20ft 0in long over headstocks, and had a 12ft 0in wheelbase. Some, if not all, built in the years 1924–6, incorporated typical Midland features of racks on each side of the body, into which could be fitted boards with letters painted thereon, which gave in code form a description of the train, its originating yard, and its destination, see plate 5A and fig 4. A diagram for similar fitted vehicles, D1658 was issued, but, although such vehicles existed, all were built to Midland lots. The LMS lots for this diagram were cancelled and re-issued to cover vehicles to LNWR diagram 17B, these being of the well known 'Crystal Palace' type.

The first breakaway from pure Midland design occurred in 1926, when the first vehicles with guard's lookouts or duckets appeared. This feature had never appeared on any of the MR vans, this company being the only one of the major constituents of the LMS not to have incorporated them in its later designs. It was thus probable that pressure from the other divisions led to their adoption on practically all subsequent vehicles built for the LMS.

Two diagrams were issued, the first, D1656 being for fitted vehicles. Width over duckets was 8ft 6in, and apart from this feature the bodies were identical to the previous type. D1657 was issued for the equivalent unfitted type, one lot being piped. A small change in the body design took place, with the verandah end planking being flush to the outside, whereas the previous types had the planks recessed with the framing outside, see plate 5B. Official records show that one vehicle of lot 300 and one of lot 309 were built without lookouts, and were included in D1659; a photograph in our possession shows that at least one further vehicle from lot 309 also lacked lookouts, No 357845. This vehicle is of interest in that the verandah ends are of the later type, while the upper horizontal handrail is divided into two portions with a gap in the middle where the lookouts would be if fitted. This vehicle is thus a hybrid and it is possible that there were other examples of this type.

Before going on to the longer wheelbase vehicles, one all-steel van must be considered. It has been said that this vehicle was built in connection with the all-steel coaches supplied by outside contractors to the LMS in the mid twenties. Since the builders plate shows it to have been built at Derby in 1926, however, there is some doubt about the validity of this suggestion. The vehicle is illustrated in plate 5C.

The next development concerned the provision of longer wheelbase vans, in order to obtain better riding at the higher speeds then being attained by some goods trains. The first of these vehicles to D1940, was built in 1933; it retained the 20ft 0in length over headstocks, but had the wheelbase increased to 14ft 0in. The body was almost identical to D1657, and the vehicles appear to have been an interim measure, see plate 5D, for in the same year the first of the true long wheelbase vehicles appeared.

The next batch was 24ft 0in over headstocks, and had a 16ft 0in wheelbase. As the body length remained practically the same as on the earlier vehicles this meant that the verandahs were considerably longer. The first diagram D1890 had wooden verandah ends, and to maintain the access openings at approximately the same width, short half height sides were fitted at the outer ends. On this and subsequent construction the planking was changed from vertical to horizontal. The diagram indicates a width over duckets of 8ft 6in, although photographic evidence shows that some were 9ft 0in wide, these probably being later replacement fittings. The livery of these vehicles when built was identical to that shown on the earlier types, so we have chosen to illustrate this type with a photograph of one in departmental stock in 1965. It also shows the V-strapping which was fitted to many of the brake vans built to this and all of the other diagrams mentioned so far, but which was never found to be the best of our knowledge on any of the later diagrams, see plate 6A and fig 5.

All subsequent construction in this group had the verandah arrangement reversed, that is the half height sides formed an extension to the body sides, the access openings thus being at the outer ends. The design of the verandah end was also modified to incorporate steel panelling on a wooden framing.

Diagrams 1919 and 2036 were identical apart from the width over duckets, being quoted as 8ft 6in and 9ft 0in respectively. On these vehicles visible weights, slung between the underframe members, first made their appearance, and on these two diagrams were quite shallow, see plate 6B.

Diagram 2068 was almost identical to D2036 apart from the weights which were much deeper, and which nearly obliterated the daylight between the underframes and step-boards. Vehicles to this diagram were built well into British Railways days, and appear in the BR diagram book with numbers in the B series.

Vehicles built to the final diagram in this group, D2155, were unique among the later LMS vans in that they lacked lookouts. The verandah end was glazed at one end only, and thus became a vestibule, while at the same end tubular steel gates were provided across the access openings in place of the customary safety bar. All of the other vans described so far had a vertical brake column mounted within the body, but on these vehicles the hand brake wheels were mounted horizontally on the ends of the body; long shafts were attached to them and worked through bevel gears to a vertically mounted screw within the body. These vehicles were almost certainly built for a special working, and there may be some connection with the Wapping tunnel (Liverpool) vans described later in this chapter; they thus form a link between the normal traffic vehicles and those built for a special purpose.

The first of the special purpose brake vans were built in 1930 to D1799. They were bogie vehicles having a tare weight of 40 tons, and were the heaviest to be built for a British railway. They each replaced two 20-ton ex-LNWR vehicles on the line between Copley Hill (Leeds) and Armley. They were 30ft 0in over headstocks, and were mounted on two four wheeled bogies of 5ft 6in wheelbase, set at 19ft 0in centres. The body maintained the standard length inside, long half height extensions filling in the excessive length of the verandah; they thus anticipated the design of the later standard vehicles. A feature not found on the normal brake vans was the provision of sanding gear, operating on the inner pair of wheels on each bogie. Apart from their large size, however, they were in outline typical of LMS practice, see plate 6C.

Next to be considered are the vehicles built to D2096 which appeared in 1941, and were built to Railway Clearing House standards. The object was to provide a vehicle acceptable to all four main line companies, for although brake vans in general did not carry non-common user markings, they were in fact always treated as such. This became apparent when the first BR standard vans appeared, branded with the region to which they belonged. The design was reminiscent of LNER practice, which in modified form became the BR standard type. The vans were 24ft 0in long over headstocks on a 16ft 0in wheelbase. The body was mounted centrally, being 10ft 0in long inside, and 17ft 6in over verandahs; prominent sand boxes were

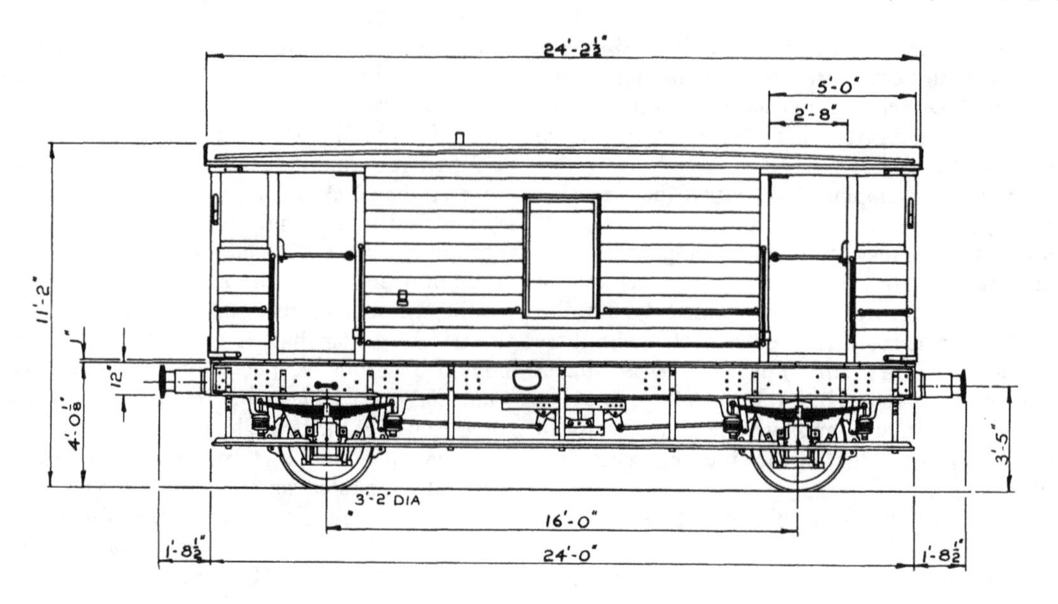

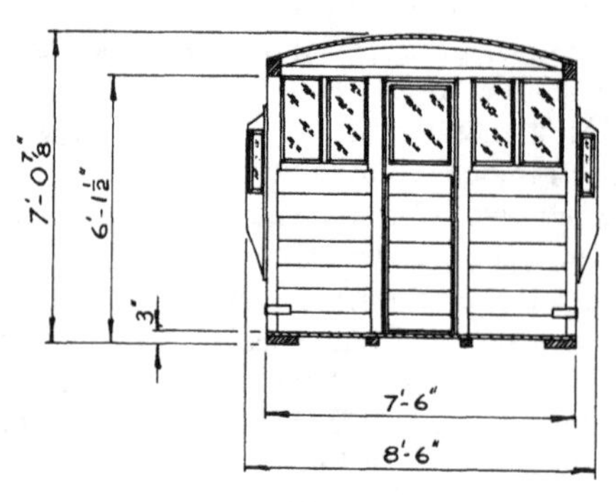

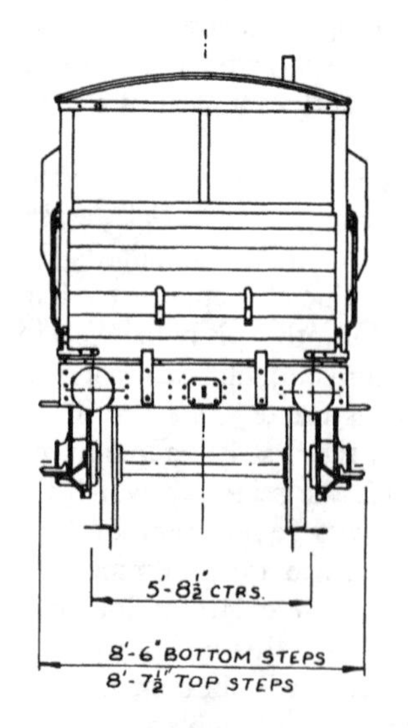

Fig 5 D1890 Goods brake van

mounted at each end of the underframe, with a sand pipe on the outer side of each wheel. Other main dimensions were as for the standard LMS vehicles, although the dimension over lookouts was a compromise at 8ft 9in. Another small point concerned the tare weight which, although nominally 20 tons, was in fact different for each of the four vehicles concerned, see plate 6D.

Finally in this group come the vehicles to D2148. Although built after nationalisation they were a throwback to pre-grouping practice, being six-wheeled vehicles. They were 18ft 0in over head-stocks, total wheelbase 10ft 0in, and had a tare weight of 20 tons. In outline they were similar to the long-wheelbase vehicles, the body being only 7ft 5in long inside. They had many features in common with the vans to D2155 already described, having the same vestibule with gates at one end only, they also lacked lookouts and had the same arrangement of brake handwheels. In addition they had sanding gear at the vestibule end only. They were built for working through the Wapping tunnel in the Liverpool dock area, see plate 7A.

Plough Brake Vans

THE last vehicles in this chapter are the plough brake vans, sometimes also described as ballast brake vans. These vehicles belonged to the civil engineer's department, and as the name implies were used on ballast trains. They incorporated a

Plate 6A, *top*, shows the final condition of the first Stanier design of goods brake van in departmental use with strengthening 'V' strapping. As far as is known this modification was of British Railways origin and was not fitted by the LMS. Most but not all vans in their final years were so altered. R. J. Essery. Plate 6B, *centre upper*, is an example of the later Stanier design in its first guise, ex-works May 1938. Later vehicles, of this type, had the extra weight in the form of infilling visible below the solebars whereas the first examples only had it below the floor between the solebars. Plate 6C, *centre lower*, illustrates one of the 40T special goods brake vans which spent most of their life on the Copley Hill-Armley duty and is in ex-works, as built condition. Another ex-works van is one of the four examples built to D2096 and illustrated as plate 6D, *bottom*, and more fully discussed within the chapter.

Plate 7A, *top,* illustrates one of the goods brake vans built to an LMS design by British Railways for the Wapping Tunnel (Liverpool) working. Plate 7B, *centre,* is an example of a plough brake van to D1805 in its original ex-works condition. Finally Plate 7C, *bottom,* shows the interior of an early design of goods brake van, probably a vehicle from D1659.

plough at each end which could be lowered as required in order to spread the ballast dropped from hopper wagons when re-laying track.

The first two vehicles, to lot 372, may or may not have fallen into this category. They were shown as belonging to D1657, but this was then crossed out and the word 'plough' substituted; no further details are available and they must, therefore, remain one of the minor mysteries of the LMS stock.

The two diagrams which were issued for plough vans shared principal dimensions. Tare weight was 16 tons and they were 21ft 0in over headstocks. Wheelbase was only 9ft 0in in order to accommodate the ploughs which were mounted beneath the underframe. The points of the ploughs projected beyond the underframe, and it was probably because of this that buffers 1ft 8½in long were used although the vehicles were unfitted. Body style was reminiscent of the long wheelbase vehicles to D1890, the body being 8ft 6in long inside. A prominent feature on the verandahs were the large handwheels used to raise and lower the ploughs. The first vehicles to D1805 were built in 1932, while the second diagram D2025 was for one vehicle only built in 1939. The only apparent difference was that D1805 was 8ft 6in wide over lookouts, while D2025 was 9ft 0in wide, see plate 7B. One further vehicle of this type was built to an LMS lot number from a St Rollox drawing; little is known of this vehicle, although it was undoubtedly of the same general type.

Livery Details

The first livery employed on these vehicles was reminiscent of Midland style. Company initials were 18in high, and were placed just above the horizontal handrails on the body-side, the M being central longitudinally. The running number was normally painted in 5½in high characters on a black panel, as described in chapter two. This was placed above the M about 12in below the roof line; cases are known, however, where the black background was dispensed with, the number being painted directly on to the grey paint with a white border of approximately the same shape and dimensions as that normally provided, an example of this being 357845.

The only other marking on the body was the tare weight, and this was at first shown in full, viz, 20 Tons, and was placed on the lower edge of the body under the L, see plates 5A and 5C. With the introduction of the vehicles with lookouts, this scheme was modified slightly. The company initials were placed somewhat higher on the body-side, the M being placed on the lookout, with its lower edge just above the lower edge of the vertical portion of the latter. The number panel was also placed somewhat higher so that it cleared the lookout. It also appears that, with the introduction of these vehicles, the tare weight marking in the majority of cases was abbreviated to 20T, in some instances also being moved further to the left end of the vehicle.

These schemes seem to have been used on all of the vehicles up to and including the first long wheelbase type to D1890. The first lot and at least some of the second lot of D1919 also appear to have been turned out in grey livery, with large company initials as before. On these vehicles the running number was painted in 4in high characters at the left hand end of the body, just below the top horizontal handrail, the black panel being abandoned; at the same time the tare weight marking was moved on some vehicles to the new 'correct' position at the right hand end of the body.

With the introduction of the bauxite livery, the standard insignia became the rule. In the majority of cases the correct positions appear to have been used; as these vehicles had no carrying capacity, however, the company initials were placed directly above the running number. In some cases the tare weight marking was moved into the position normally occupied by the carrying capacity, and here the vertical positions varied considerably. Some vehicles had the company initials placed just above the top handrail, tare weight and running number being placed between the handrails, while in other cases the initials and tare weight were both above the handrails. Where the correct positions were used the company initials were placed between the handrails, with the running number and tare weight at the lower edge of the body.

With the advent of nationalisation, company initials disappeared, the running number being prefixed M. This was placed at the lower left hand end of the body, tare weight being placed either at the right hand end, or in some cases above the running number.

The vehicles built to D2068 with numbers in the B series had the running number at the left hand end, placed just above the lower handrail with the words London Midland placed immediately below it, tare weight being at the right hand end.

The vans built for special duties followed the schemes as outlined above, depending on the date of building, but had in addition wording describing the duty for which they were built, see plates 6C and 7A.

The plough brake vans also followed the scheme, but in addition had letters to describe their function and area allocation. These consisted of the letter E at the left hand end; this was about 12in high and denoted Engineers Department, while at the right hand end a letter or letters denoted the area. This was again about 12in high where a single letter was employed, ie W=Western Division, while where multiple letters were employed they had a total height of about 12in, S/SW=Scottish South Western. It would appear that they were not reduced in height with the change in livery and style. However, there is some evidence to suggest that the original grey was darker than that used on normal freight stock.

Summary of Brake Vans

20ft 0in OVER HEADSTOCKS 12ft 0in WHEELBASE TARE 20 tons

Diag	*Lot*	*Qty*	*Built*	*Year*	*Numbers*	*Remarks*
1659	37	150	Derby	1924/5		Code BRO
	114	100	,,	1926		
	Pt119	97	,,	1924/6	Various	
	136	100	,,	1925		
	199	100	Stableford	1925/6	357401–357500	
	200	100	B'ham C & W	1925	357501–357600	
	201	100	Clayton	1925	357601–357700	
	241	100	Derby	1926	Sample 281304	
	Pt300	1	,	1927	357711	
	Pt309	1	,,	1927	357885	
Sample Numbers 134900, 144182, 280663, 294048						
All-Steel Possibly built as part of lot 241				Number 281304		
1658	63	70	Earlstown	1924		LNWR Diagram
	Pt119	3	,,	1924	Various	17B see text
1656	242	100	Derby	1926/7		Fitted Code BRO
Sample Numbers 277, 542, 2045 (D1656)						

Diag	*Lot*	*Qty*	*Built*	*Year*	*Numbers*	*Remarks*
1657	Pt300	99	Derby	1927	357701–357710	Code BRO
					357712–357800	
	Pt309	99	,,	1927	357801–357884	357845
					357886–357900	without
	316	50	,,	1927	357901–357950	lookouts
	372	2	Leeds Forge	1928		Lot 372
	374*	200	Derby	1928/9		Plough?
	453	100	,,	1929	Various	
	503	100	,,	1930		
	540	250	,,	1930/1		
	541	50	,,	1931		Piped

*Alternative records suggest these vehicles may have been built at Newton Heath
Sample Numbers 284816, 284943, 284990, 284992, 287329, 291834, 294267, 295220, 296257, 296602, 299603, 318643, 319173, 319463, 319665, 320334, 325063, 326166, 328337, 329374.

20ft 0in OVER HEADSTOCKS 14ft 0in WHEELBASE TARE 20 tons

Diag	*Lot*	*Qty*	*Built*	*Year*	*Remarks*
1940	671	100	Derby	1933	Code BRO
Sample Numbers 295384, 295516, 297435.					

24ft 0in OVER HEADSTOCKS 16ft 0in WHEELBASE TARE 20 tons

Diag	*Lot*	*Qty*	*Built*	*Year*	*Numbers*	*Remarks*
1890	715	50	Derby	1933		Code BRO
	716	50	,,	1933	Various	
	757	50	,,	1934		Piped
Sample Numbers 286163, 294346, 294350, 295083, 295443, 295661, 295721, 296312, 296412, PIPED 296406, 296622, 297699, 296708.						
1919	835	90	Derby	1935	730000–730089	Piped
	919	120	,,	1936	730090–730209	Code BRO
	1007	160	,,	1937	730210–730369	
	1103	150	,,	1938	730370–730519	
	1104	150	,,	1938	730520–730669	
2036	1204	72	Derby	1940	730670–730741	Code BRO
	1205	72	,,	1940	730742–730813	
	1278	75	,,	1940	730814–730888	
	1279	75	,,	1940	730889–730963	
	1281	100	,,	1940	730964–731063	
	1291	128	,,	1941	731064–731191	

Diag	Lot	Qty	Built	Year	Numbers	Remarks
2068	1316	100	D'by & W'tn	1942	731192–731291	Code BRO
	1332	250	,,	1942/3	731292–731541	
	1343	200	,,	1943/4	731542–731741	
	1363	150	Wolverton	1944	731746–731895	
	1387	191	D'by & W'tn	1945	731896–732086	
	1424	150	Derby	1946	732096–732245	
	1470	66	,,	1948	732246–732311	
	1471	75	,,	1947/8	732321–732395	
	1517	75	,,	1948/9	732396–732470	Fitted
	1518	75	,,	1948/9	732471–732545	
2155	1411	9	Wolverton	1947	732087–732095	

BOGIE BRAKE VANS TARE 40 tons

Diag	Lot	Qty	Built	Year	Numbers
1799	549	3	Newton Heath	1930	284723–284725

24ft 0in OVER HEADSTOCKS 16ft 0in WHEELBASE RCH SPECIFICATION

Diag	Lot	Qty	Built	Year	Numbers	Tare
2096	1352	4	Wolverton	1941	LMS 731742	21t 6c 2q
					LNE 760948	20t 6c 2q
					GWR 35927	20t 10c 2q
					SR 56060	20t 11c 2q

Alloted LMS numbers 731742–731745, numbers carried as shown.

18ft 0in OVER HEADSTOCKS 10ft 0in WHEELBASE (SIX WHEELS) TARE 20 tons'

Diag	Lot	Qty	Built	Year	Numbers
2148	1498	9	Derby	1948	M732312–732320

BALLAST & PLOUGH BRAKE VANS
21ft OVER HEADSTOCKS 9ft 0in WHEELBASE TARE 16 tons'

Diag	Lot	Qty	Built	Year	Numbers
1805	635	9	U/F Cravens Body Derby	1932	197263–197271
2025	1161	1	Pickering	1939	748700

DIMENSIONS UNKNOWN BUILT TO St ROLLOX DRAWING 11328

Diag	Lot	Qty	Built	Year
	225	1	St Rollox	1927

CHAPTER 4

OPEN GOODS WAGONS

THE open goods wagon was, without doubt, the mainstay of the wagon fleet. This is shown by the fact that, of the 305,000 wagons taken over by the LMS at the time of the grouping, no fewer than 162,000 came into this category; at the same time the covered goods vans totalled less than 40,000. Even at the time of nationalisation the open goods wagons were twice as numerous as the covered variety.

The vehicles taken over by the LMS were of many different designs, and many thousands of them were relatively elderly, having a carrying capacity of 10 tons or less. The LMS set about replacing them with new vehicles having a carrying capacity of 12 tons. This policy was pursued with such vigour that 62,650 new vehicles of this type were built in the years 1923–30. These vehicles carried a wide range of general merchandise, the exceptions being perishables and articles likely to suffer from the effects of exposure to atmosphere. Bulky, though light, loads could easily be loaded into these wagons, which it would have been difficult if not impossible to load through the restricted door opening of a covered goods van.

Open goods wagons were divided into three broad groups—high-sided, medium and low-sided goods respectively. The high-sided version generally had a body formed from five planks, commonly referred to by enthusiasts as 'five plank opens'. Railwaymen however were more interested in the height of the sides, hence the official classification. They had a door in each side, usually about 5ft wide, and extending over the whole height of the side, thus giving easy access to the interior when open.

Tarpaulin sheets were widely used, and some companies provided tarpaulin bars, which ran the length of the vehicle and could be lowered to one side when not required; as far as can be ascertained none of the LMS standard vehicles were built with,

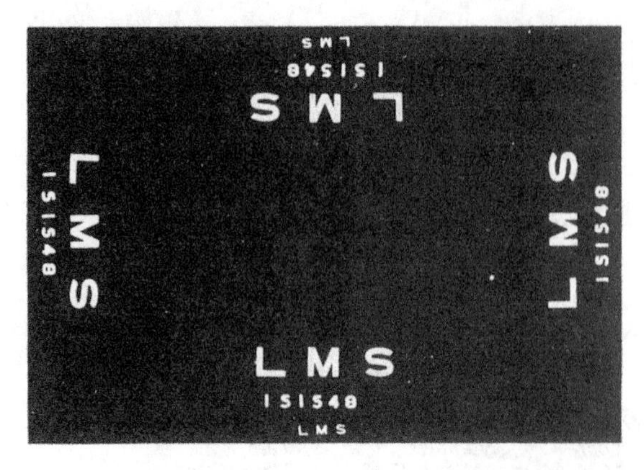

Left: Fig 6 Wagon tarpaulin layout; in some cases numbers were up to 50 per cent larger.

Below: Plate 8A depicts a typical open goods wagon of D1666 in photographic grey livery—unusual for a goods vehicle. Note the grey W irons, brake gear etc. In ex-works condition, as far as black paintwork is concerned, refer to Plate 8D; 158125 was photographed on 16.1.1929.

or acquired, this feature during the LMS era, although some of them were so fitted under BR.

The tarpaulin sheets themselves were made of grey canvas weighing 12 to 16oz/sq yd. This was supplied in bales of about 100yd in length, and 36 or 42in in width. This was cut to length, and sewn together to make a sheet 21ft long by 14ft 4in wide. During the sewing, strengthening tabs were inserted, to take the cords used for securing the sheet to the wagon. Holes were then punched into the tabs, and iron rings sewn in for additional strength. Sheets were then taken to the dressing shop, where the waterproof dressing was applied. Three coats of the dressing were applied, the sheets being hung up to dry between each coat. The sheets were lettered and numbered, the securing cords placed through the eyebolts, and the sheets again hung to dry until required for traffic. Average life of a sheet was about five years, see fig 6.

Medium goods wagons generally had bodies formed from three planks, and on these vehicles the whole side was arranged to drop down. The low-sided version had sides consisting of one plank only, generally fixed, so that loads could easily be lifted over them.

A development of the open goods wagon was to be found in the shock-absorbing wagons developed in the late 1930s, the idea being to minimise damage to fragile goods during shunting operations. Both high and medium types were produced, while latterly the principle has also been extended to covered vehicles.

In Plate 8B, to illustrate D1667, the less numerous steel-framed version is a vehicle in BR condition, as photographed in 1964 at the end of its useful life. This vehicle was built by the Butterley Company in 1926. Note the split spoke wheels at the left hand and solid spoked wheels at the far end. Repairs have caused the original axleboxes to be replaced by axleboxes of LNER pattern. R. J. Essery

High-Sided Goods

HIGH-SIDED goods wagons, sometimes described as merchandise wagons, were probably the most familiar types in the pre-nationalisation goods train. Unlike brake vans, more than one type of construction was in use concurrently; strict chronological order of construction is thus made more difficult

Body construction was relatively simple, and was composed mainly of timber about 7in wide by 2½in thick. The floor consisted of transverse timbers fastened directly to the underframe, the horizontal body timbers being secured by angle irons, corner plates, tee-section end stanchions and flat-section strapping.

Before describing standard LMS vehicles, mention must be made of lot 39. This was for vehicles of 10 tons capacity, of LNWR design; these were probably of the all-wood type, and are one of the few examples of pre-grouping designs built to LMS lots.

All the remaining vehicles in this section were of 12 tons capacity, most being uprated to 13 tons during the war years. Cubic capacity was also quoted on the diagrams, and varied from 408 to 427 cubic feet. All were built on standard underframes, width over body being 8ft 0in. The height of the sides varied from about 3ft 1in to 3ft 3in.

The first diagram to be issued was D1666. This was a Midland Railway design, and was at first given D1338 in the MR series, the first lot built in 1923 being included in the MR lot book. These wagons were of the all-wood type, and were of completely conventional design, wheelbase being 9ft 0in. They proved to be far and away the most popular when it came to quantities built, no fewer than 54,450 being constructed in the years 1923–30. As may be expected with such a large class, numerous small variations in detail occurred, while the running numbers extended over a very wide range, no block numbering being encountered. All these vehicles were built in the company's own workshops, Derby in particular being very prominent. Perusal of the summary table shows that this works was building about 7000 to 8000 vehicles of this type alone every year for several years. Since this works out at about 150 vehicles a week, it would seem that the railways had little to learn about mass production methods; see plate 8A.

Contemporary with these vehicles were those built to D1667. This design was virtually identical as regards the body, but was mounted on a steel underframe. Quantities involved were much smaller, and it will be noted that except for the last two lots all were built by outside contractors, see plate 8B and fig 7. All of these vehicles were unfitted, and were provided with double hand brakes.

In common with other types of vehicle, the next development was an increase in wheelbase, in this case to 10ft 0in. Three types of construction are encountered, and as the building dates overlap considerably they will be considered in groups.

The first were of the all-wood type, D1895 being very similar in appearance to D1666. The only major difference, apart from wheelbase, occurred in the doors, which on this and all subsequent construction were of the so-called 'merchandise' type. These doors had the bottom plank angled outwards, the reason for this being to give a smaller gap between the door and wagon floor

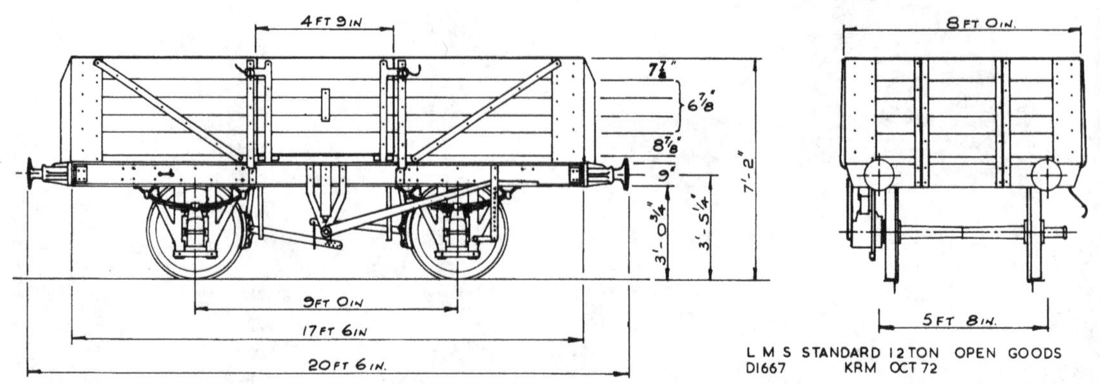

Above: Fig 7 D1667 Merchandise wagon

Below: Fig 8 D1892 Merchandise wagon

Right: Plate 8C This was the only vehicle built to lot 810 at Derby in 1934 and was allocated to D1892. A. E. West

when the former was resting on a loading dock bank, thus making for a smoother exit for the barrows often used for loading and unloading these vehicles.

The second diagram in this group, D1896, was for very similar vehicles, but built without curb rails (see fig 1). All the vehicles in this group were unfitted, and, with the exception of lot 808 which had the double type, were provided with Morton type handbrakes. Although relatively few in number, construction was spread over the years 1934–9 and there is thus a considerable overlap with the other types of construction.

The final diagram in this group, D2073, is of interest in that the wagons could never have run on the parent system. These vehicles were built for the Northern Counties Committee (Ireland), and as such were for the 5ft 3in gauge employed on that line. They were of conventional appearance, the bodies being somewhat wider at 8ft 7in, while the capacity was 457 cubic feet. It is also of interest to note that they were built for the LMS by the LNER; they were unfitted and provided with Morton hand brakes. The date of building is not certain but was about 1942.

The second group covers the vehicles with all-wood bodies and steel underframes. Four diagrams were involved, D1892, 2072, 2094 and 2151. All were similar in appearance, the reason for the various diagrams being that internal dimensions varied according to the thickness of timber used in their construction, while the internal height of the sides also varied slightly. Construction was spread over the years 1934–46, although D2151 which were built for the Ministry of Supply, probably during the war years, were not shown as taken into stock until 1949.

D1892 included both unfitted and fitted examples, the latter having a modified form of Morton brake with two brake shoes per wheel, and shorter side levers. All unfitted vehicles in this group had the standard Morton type of brake

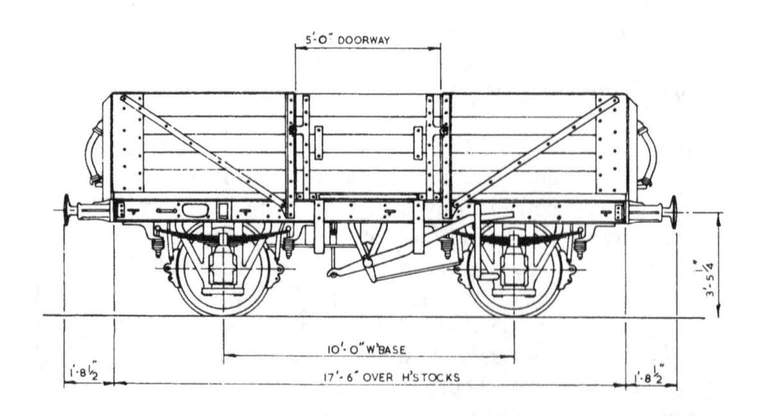

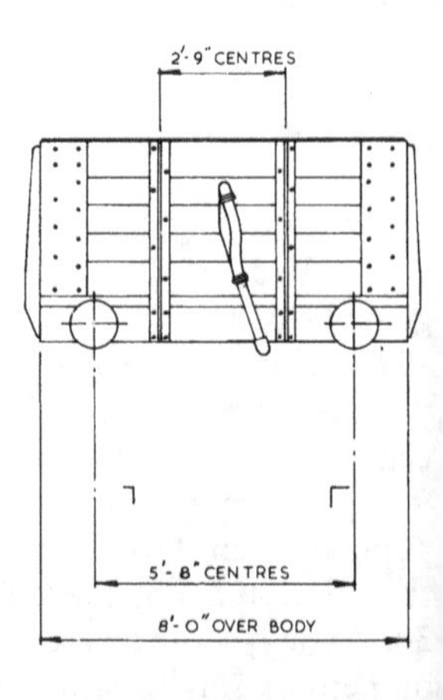

gear with the exception of D2151 which had the double type. To illustrate this group we have chosen a fitted example from D1892. (Plate 8C and fig 8).

Two interesting modifications to D1892 changed the classification for they were no longer regarded as open goods. The first modification, to D2047, consisted of the provision of racks to hold oxygen or hydrogen cylinders. and the fitting of a shallow curved roof. These vehicles thus had the appearance of a very low covered van. The second modification was allocated D2057; a much simpler conversion, consisting of the provision of fixed wooden cradles, designed to carry crated aircraft propellers. Both were wartime measures, and it is possible that the vehicles concerned reverted to their original condition later, since no major structural changes were involved.

The third group covers those vehicles with steel underframes and wooden bodies having corrugated steel ends. The first vehicles of this type appeared in 1933; there was then a considerable gap, construction recommencing in 1946. Three diagrams in all were issued, D1839, 2110 and 2150; again there is little to distinguish the three types, the main variations being in the internal dimensions. The steel ends were wood lined, and were of a different type from those used on the covered goods. On these vehicles the corrugations were pressed inwards, while on the covered vehicles they were pressed outwards, this being very apparent in photographs. All the vehicles to D1839 were fitted, and had the Morton type of brake gear with one brake shoe per wheel. D2110 and 2150 included both unfitted and fitted examples, in this case the brake gear was the same as that already described for the vehicles to D1892, see plate 8D.

Medium Goods

In direct contrast to the high-sided goods, where a very large replacement programme was put in hand immediately after grouping, the LMS appears to have been quite satisfied with its inherited stock of medium goods, for the first standard vehicles did not appear until 1935 (see fig 9).

As a direct consequence, there is not the variety of designs that occurred with the high-sided vehicles. Only two standard diagrams were involved, and as they were practically identical can be considered together.

D1927 and 2101 had all-wood bodies, mounted on standard steel underframes with a 10ft 0in wheelbase. D1927 was rated to carry 12 tons, while D2101 wagons were rated to carry 13 tons when built in 1945, capacity being quoted as 223 cubic feet. Both were 8ft 0in wide over the body, thinner side planks on D2101 making the internal widths 7ft 7in and 7ft $9\frac{1}{4}$in respectively. The whole of the bodyside was arranged to drop down. To counterbalance the weight, sprung levers were employed, the final lever in the system being prominent in the middle of the sides, see plate 9A. Some vehicles to D1927 were fitted, the brake gear having two blocks per wheel, with short side levers to actuate the hand brake. The unfitted vehicles of both diagrams had the normal Morton type of handbrake.

A third diagram, D2147 was issued to cover some ex-Ministry of Supply vehicles; they were of all-steel construction, and had small side doors only. They were 16ft 6in long over headstocks and had a

LMS
12 T.
6·13.0
N

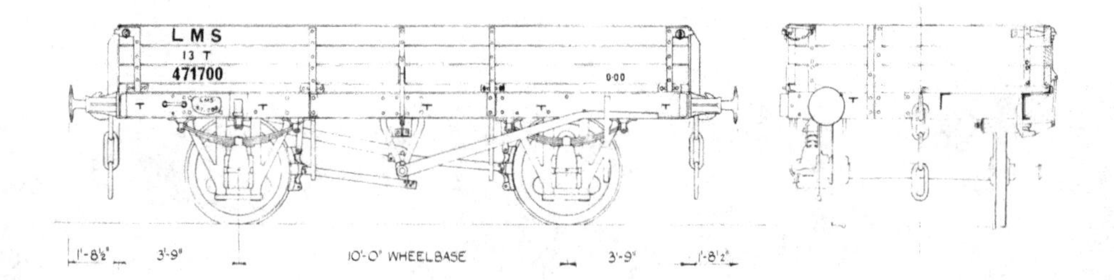

Above: Fig 9 D1927 Medium goods wagon

9ft 0in wheelbase; they were unfitted and had the double type of hand brake, see plate 9B.

Low-Sided Goods

THE LMS appears to have been well provided with this type of vehicle also, for it was not until 1938 that the vehicles to D1986 appeared. Even then only 1000 of the type were built, which emphasises their relatively restricted use. Their low sides precluded loading with loose articles, and they were thus used mainly for crated loads, containers, or as runners for bolster wagons. Construction was of wood and the body, with fixed sides, was mounted on a standard steel underframe having a 10ft 0in wheelbase. They were unfitted, and were provided with the Morton type of hand brake, see plate 9C and fig 10.

Shock Absorbing Wagons

DURING the 1930s, considerable concern was shown by the railways over damage to fragile goods during shunting operations. To help overcome this problem, the shock absorbing vehicle was developed. The basic principle was simply to allow the body to slide longitudinally

Top left: Plate 8D illustrates an example of D1839 in ex-works condition. In particular, attention should be drawn to the positions of the tare weight on 3466, and its absence on 400231 and 158125.

Centre left: Plate 9A is an example of D1927 as built with bauxite livery, while plate 9B, *bottom left,* illustrates a vehicle from D2147 in ex-works condition.

Below: Fig 10 D1986 Low-sided goods wagon

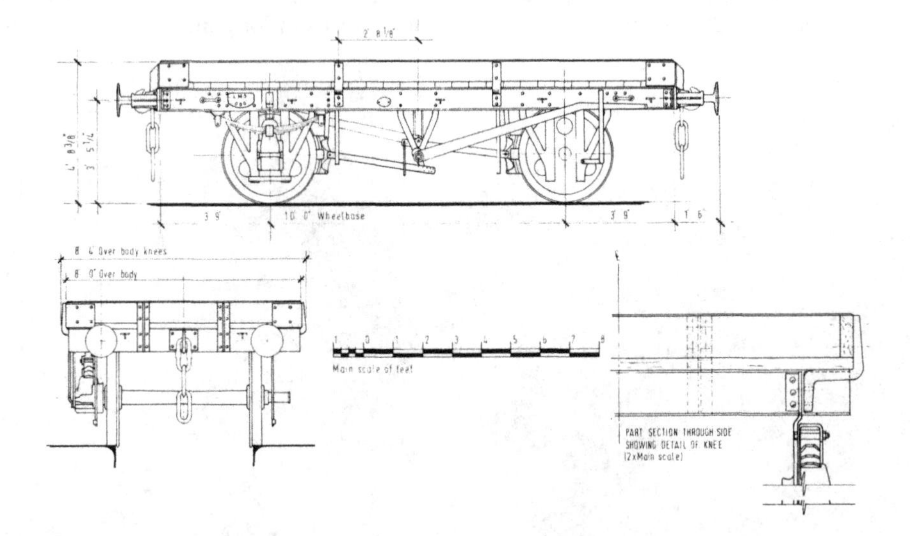

under the control of springs. Thus when the vehicle hit another during shunting, or was otherwise stopped suddenly, the body slid along on the underframe, compressing the springs, and this had the effect of reducing the rate of deceleration for the load.

The first experimental vehicles were converted very cheaply, using existing material. The bodies were originally built as roll-off containers, of all steel construction, and described in chapter 14. No lot numbers were allocated for the conversion, and little major modification appears to have been required. In both container and shock absorbing guises, standard steel underframes with a 9ft 0in wheelbase were employed, and the conversion involved the substitution of longitudinal spring mountings for the fixed type. Carrying capacity was quoted as 10 tons or 273 cubic feet. Six were so converted in 1935, and while the underframes retained their original running numbers, the bodies retained the original container number but prefixed SAW instead of ROC, see plate 26A.

The first of the shock absorbing vehicles built new was to D1979, this being for one vehicle only. It had a standard steel underframe with a 10ft 0in wheelbase. The body was of the five plank high goods type, with corrugated ends, it was 16ft 11⅜in long outside and 16ft 6in inside, no wood lining being fitted in the ends. D1983 followed (figs 11 and 11A), and differed only in having a wood lining to the ends, which reduced the inside length to 16ft 3¼in. These two diagrams were unfitted and were provided with the Morton hand brake, while the final diagram in this group, D2040, was identical to D1983 apart from the provision of vacuum brakes. These vehicles had two brake shoes per wheel, the hand brake levers being of the normal long type. All, when built, had exposed control springs, later fitted with cover plates as a safety measure, the carrying capacity being 12 tons in all cases.

A medium goods variant was produced, to D2152 appearing in 1949. They had an all-wood body 16ft 6in long outside; this was of the three

Left: Plate 9C shows a new vehicle to D1986 in service as a runner for an overhanging load, photographed at Bedford on 29 November 1938, while plate 9D, *bottom left,* is an example of D2152, an LMS shock absorbing design built by and finished in British Railways livery at the time of its construction.

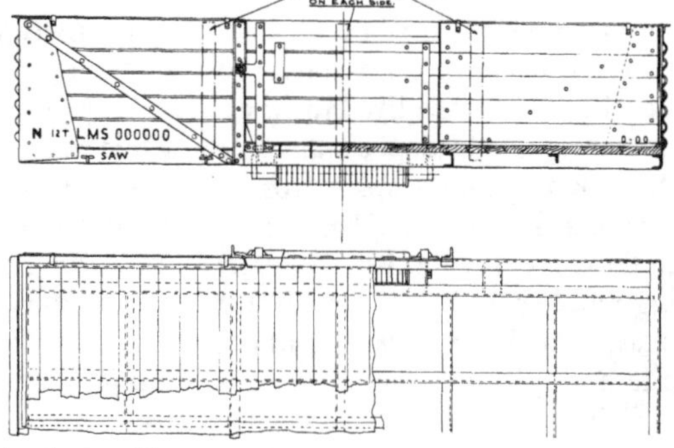

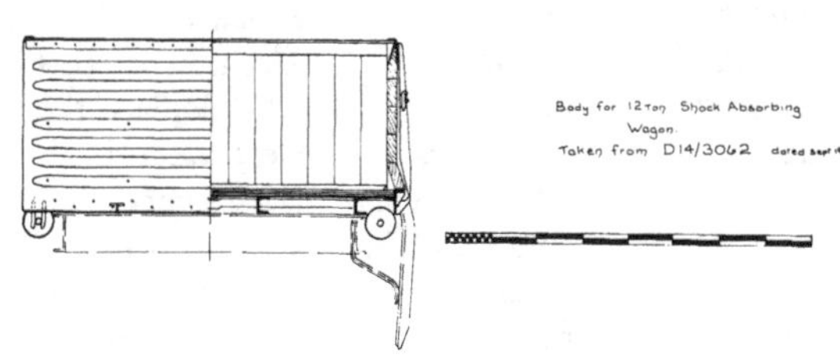

Right and below: Fig 11 & 11A D1983 Shock absorbing wagon, body and underframe.

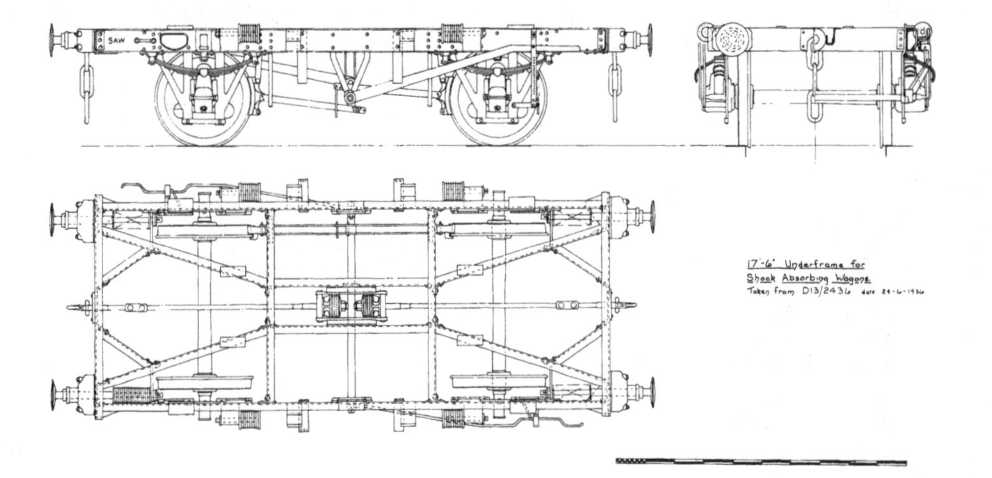

plank type and the sides were arranged to drop down as on the conventional vehicles. They were built specifically for glass traffic, and were fully fitted, the hand brake again being of the Morton type. Carrying capacity was 13 tons, see plate 9D.

Livery Details

The high sided goods wagons at first had large company initials, and although they varied in size, particularly on those vehicles built by the trade, they were in general about 18in high. They were placed on the middle three planks of the body, and were usually placed to avoid the strapping.

The running number was painted on the bottom plank, at the left hand end, clear of the corner plate, while the tare weight which was given in figures only for tons, cwts and qtrs was placed in one of three main positions. The most common position was on the bottom plank (or curb rail where fitted) at the right hand end; alternately it was placed in the same relative position at the left hand end, this position being popular on vehicles built by the trade. The final position was on the solebar to the left of the right hand wheel, but this appears to have been much more unusual. This scheme was applied to all unfitted vehicles up to about lot 964; the fitted vehicles to D1892 in addition had the X denoting their status painted on the two bottom planks at the right hand end just clear of the corner plates. They also had the non-common-user markings painted at each end on the corner plates in line with the bottom plank.

With the change to bauxite livery, the correct positions as laid down in Chapter Two appear to have been followed almost universally. However, those vehicles built during and after the war came out with the woodwork unpainted apart from a patch on the bottom plank at the left hand end; on this was painted the carrying capacity, company initials and running number in that order; in this case the lettering started at the extreme end of the vehicle, while the tare weight was painted on the corner plate at the right hand end.

The first lots of the medium goods to D1927 also had the large company initials, and in this case they extended over all three planks; the majority of these vehicles, however, were built after the introduction of the bauxite livery, and appear to have followed the standard scheme, although the application of non-common-user markings to the fitted examples does not appear to have been universal.

All the low-sided goods had bauxite livery when built, and as the sides were not high enough for the normal positioning of the insignia, they were placed in line as follows: the running number, carrying capacity and company initials were placed at the left hand end in that order, while the tare weight took the right hand end.

An example of the livery used on the experimental shock absorbing wagons is shown in plate 26A. As already explained these vehicles were conversions of containers, and further livery details will be found in chapter 14. The only point to be noted here is that although the two parts were no longer seperable they were still given their own running numbers.

The shock absorbing wagons which were built as such had bauxite livery from the first, and had the insignia in the correct positions; the code SAW was painted on the body below the other insignia at the left hand end, and all appear to have had non-common-user markings, At first these vehicles carried no other distinguishing marks, but later had the three vertical white stripes painted on the body sides to denote their status, see fig 3; this was probably applied to vehicles from 450100 on when built.

Summary of Open Goods Wagons

High Sided Goods

10 ton OPEN WAGON NO DIAGRAM
LNWR DESIGN
Lot 39 Qty 250 Built Earlstown 1924
12 ton CAPACITY ALL WOOD TYPE
9ft 0in WHEELBASE D1666
Tare 6 tons 19 cwt Code HGK Fitted Double Brake.

Lot	*Qty*	*Built*	*Year*
1005*	1000	Derby	1923
6	1000	"	1923/4
17	1000	"	1924
18	1000	Newton Heath	1924
19	1000	Earlstown	1924
67	1000	Derby	1924
68	1000	"	1924
84	1000	"	1924
89	2000	"	1924
92	750	"	1924
110	1000	Newton Heath	1924/5
116	1000	Derby	1924/5
117	1000	Newton Heath	1925
140	2000	Derby	1925
151	2000	"	1925
205	2000	"	1925
221	2000	"	1925
226	1800	"	1925/6
240	2000	"	1926
296	2000	"	1926/7
301	2000	"	1927

Lot	*Qty*	*Built*	*Year*
323	2000	Derby	1927
339	2000	"	1927
350	2000	"	1927
362	2000	"	1928
376	1000	"	1928
393	2000	"	1928
413	2000	"	1928/9
426	1000	"	1929
427	900	Earlstown	1929
438	1000	Derby	1929
439	1000	Earlstown	1929
470	1000	Derby	1929
471	1000	Earlstown	1929
482	1000	Derby	1929/30
483	500	Earlstown	1929
515	1000	Derby	1930
535	1000	"	1930
547	1000	"	1930
548	1000	Earlstown	1930
590	500	Derby	1930

*Midland Railway lot

Sample Numbers 24361, 54680, 91919, 92978, 113012, 114872, 122580, 130487, 134571, 139905, 158125, 159093, 169147, 197803, 216051, 217624, 234421, 237015, 238611, 247185, 282102, 304008, 327674, 331426, 344839, 348708, 360790.

12 ton CAPACITY WOOD BODY/STEEL UNDERFRAME 9ft 0in WHEELBASE
Tare 6 tons 14 cwt Code HGK Fitted double brake
D1667

Lot	Qty	Built	Year
56	250	Trade	1924
57	500	,,	1924
58	250	,,	1924
173	150	Gloucester	1925
174	150	B'ham C & W	1925
175	100	Cravens	1925
176	150	Hurst Nelson	1925
177	500	Metro C & W	1925
178	100	Stableford	1925/6
179	300	G R Turner	1925/6
180	250	Midland Wagon	1925
268		Trade	1926
269		,,	1926/7
270		,,	1926/7
271		,,	1926/7
272	3000	,,	1927/8
273	for lots	,,	1926/7
274	268/76 inc	,,	1926/7
275		,,	1926/7
276		,,	1926
440	1000	Newton Heath	1929
476	1500	,, ,,	1929/30

Sample Numbers 85457, 85698, 104838, 109621, 135109, 283664, 290553, 295015, 338968.

12 ton CAPACITY ALL WOOD TYPE 10ft 0in WHEELBASE
CODE HGK Fitted Morton Brake (Lot 808 Double Brake)

Diag	Lot	Qty	Built	Year	Numbers	Remarks
1895	807	400	Derby	1934	402000–402399	Tare 6t 19c
	808	800	,,	1934	402400–403199	
	1118*	800	,,	1938/9	413650–414449	
1896	809	700	Derby	1934	403200–403899	Tare 6t 10c

*This lot is shown in the diagram book as D1896, photographs and official drawings however show them to have had curbrails, which logically makes them D1895 as shown, though differing in small details from the earlier lots.

Diag	Lot	Qty	Built	Year	Numbers	Remarks
2073	1325	150	LNER		2251–2400	For NCC

12 ton CAPACITY WOOD BODY STEEL UNDERFRAME 10ft 0in W/BASE
CODE FHG (FITTED) HG (UNFITTED)

Diag	Lot	Qty	Built	Year	Numbers	Remarks
1892	783	200	Derby	1934	400000–400199	Fitted
	784	199	,,	1934	400200–400230	Fitted
					400232–400399	Fitted
	785	400	Met Cammell	1934	400400–400799	Fitted
	786	400	B'ham C & W	1934	400800–401199	Fitted
	787	400	Chas Roberts	1934	401200–401599	Fitted
	788	400	Hurst Nelson	1934	401600–401999	Fitted
	810	1	Derby	1934	400231	Fitted
	918	750	,,	1936	404000–404749	
	957	400	Hurst Nelson	1936	404750–405149	
	958	300	Pickering	1936	405150–405449	Tare
	959	300	Met Cammell	1936	405450–405749	6t 18c
	960	300	B'ham C & W	1936	405750–406049	(fitted)
	961	200	Butterley	1936	406050–406249	6t 2c
	962	200	G R Turner	1936	406250–406449	(unfitted)
	963	150	Cravens	1936	406450–406599	
	964	150	Gloucester	1936	406600–406749	
	1003	500	Wolverton	1937	406750–407249	
	1004	500	,,	1937	407250–407749	
	1024	1000	Derby	1937	407750–408749	
	1025	1050	,,	1937	408750–409779	
	1026	500	,,	1937	411250–411749	Fitted
	1030	725	Wolverton	1937	409800–410524	
	1031	725	,,	1937/8	410525–411249	
	1110	700	Derby	1938/9	411950–412649	
	1111	900	,,	1939	412650–413549	
	1119	850	Wolverton	1939	414450–415299	

2047 OXYGEN CYLINDER WAGON CONVERTED FROM D1892 IN 1940
TARE 12t 6c INCLUDING CYLINDERS Fitted 400267, 400277, 400384, 400517, 400656, 401134, 401264, 401308, 401364, 401564, 401576, 401917, 401993.
2057 AIRSCREW WAGONS IMPROVISED FROM D1892 TARE 6t 8c Unfitted

Diag	Lot	Qty	Built	Year	Numbers	Remarks
2072	1322	500	D'by & W'ton	1942/3	415300–415799	13 ton Tare 6t 6c
2094	1345	1500		1943	415800–417299	13 ton
	1353	310	D'by & W'ton	1944	417300–417609	Tare
	1371	465	,, ,,	1944	417610–418074	6t 7 cwt
	1381	750	Derby	1945	418075–418824	
	1394	750	D'by & W'ton	1946	418825–419574	
2151	1593	144		1949	M360120–263	13 ton
	1612	14		1949	M360358–71	Tare 6t 3c to 6t 12c

12 ton CAPACITY WOOD BODY CORRUGATED ENDS STEEL U/FRAME 10ft 0in WHEELBASE
CODE FHG (FITTED) HG (UNFITTED)

Diag	*Lot*	*Qty*	*Built*	*Year*	*Numbers*	*Remarks*
1839	676	100	Derby	1933	Various incl 3466	Fitted Tare 6t 17c
2110	1416	500	Derby	1946	419575–420074	Tare
	1420	2000	Wolverton	1947	420075–422074	6t 12c
	1453	300	Derby	1947	422075–422374	
	1464	250	Wolverton	1948	422375–422624	Fitted
	1473	500	„	1948	422625–423124	
2150	1510	300	Wolverton	1948	423125–423424	Fitted
	1511	1200	Derby	1949	423425–424624	Tare 7t 5c

Medium Goods

12 ton CAPACITY WOOD BODY STEEL U/FRAME 10ft 0in W/BASE
CODE FMG (FITTED) MG (UNFITTED)

Diag	*Lot*	*Qty*	*Built*	*Year*	*Numbers*	*Remarks*
1927	870	700	Met Cammell	1935	470000–470699	Fitted
	871	400	B'ham C & W	1935	470700–471099	Fitted
	872	300	Hurst Nelson	1935	471100–471399	
	873	300	Chas Roberts	1935	471400–471699	
	920	250	Derby	1936	471700–471949	Fitted
	921	260	„	1936	471950–472209	
	922	1000	„	1936	472210–473209	
	930	250	Wolverton	1936	473210–473459	
	970	240	Chas Roberts	1936	473460–473699	Fitted
	1013	800	„ „	1937	473700–474499	
	1014	650	Met Cammell	1937	474500–475149	
	1015	550	Hurst Nelson	1937	475150–475699	
	1135	150	G R Turner	1938	475700–475849	Fitted
	1136	250	Chas Roberts	1938/9	475850–476099	
	1137	250	Pickering	1938	476100–476349	
	1382	1000	Derby	1945	476900–477899	
	1417	500	„	1946	477900–478399	Fitted
	1472	250	Wolverton	1948	478400–478649	Fitted
	1521	1500	„	1949	478650–480149	Fitted
	1522	500	„	1949	480150–480649	
2101	1364	550	D'by & W'ton	1945	476350–476899	

12 ton CAPACITY ALL STEEL TYPE 16ft 6in OVER HEADSTOCKS 9ft 0in WHEELBASE

Diag	*Lot*	*Qty*	*Built*	*Year*	*Numbers*
2147	1569	4	Alt at Derby	1948	480650–480653

Low Sided Goods

12 ton CAPACITY WOOD BODY STEEL U/FRAME 10ft 0in WHEELBASE
CODE LG D1986

Diag	*Lot*	*Qty*	*Built*	*Year*	*Numbers*
1986	1113	500	Wolverton	1938	460000–460499
	1114	500	„	1938	460500–460999

Shock Absorbing Wagons

High Goods Type

12 ton CAPACITY WOOD BODY CORRUGATED ENDS STEEL U/FRAME 10ft 0in WHEELBASE
CODE SAW

Diag	*Lot*	*Qty*	*Built*	*Year*	*Numbers*	*Remarks*
1979	Pt1060	1	Derby	1937	450000	
1983	Pt1060	99	Derby	1938	450001–450099	
	1231	100	„	1939	450100–450199	
	1266	250	„	1940	450200–450449	
	1465	250	„	1949	450500–450749	
2040	1267	50	Derby	1940	450450–450499	Fitted
	1466	44	„	1949	450750–450793	Fitted

Medium Goods Type

12 ton CAPACITY ALL WOOD BODY STEEL U/FRAME 10ft 0in WHEELBASE
CODE SHOCK

Diag	*Lot*	*Qty*	*Built*	*Year*	*Numbers*
2152	1546	6	Derby	1949	489000–489005

CHAPTER 5

COVERED GOODS VANS

THE covered goods vans, which in some cases were referred to in the diagram book as ventilated vans, formed, with the open goods, the great backbone of the wagon fleet for the conveyance of general merchandise. Together they accounted for no less than two thirds of the total wagon stock taken over by the LMS at the time of grouping. Like the open goods, they were general purpose vehicles, usually being loaded with those articles which required the extra protection afforded by being totally enclosed against the elements.

The LMS issued so many diagrams for these vehicles that the reader may be forgiven if he begins to wonder, as did the authors, whether there was such a thing as a standard LMS van. Unlike the open goods all the vehicles in this chapter (with the exception of those built for the NCC) had standard steel underframes. In common with the open vehicles, the earlier examples had a 9ft 0in wheelbase, increased to 10ft 0in on the later vehicles. With the exception of vehicles built by the Southern Railway to its own design, and supplied to the LMS under a wartime arrangement, all had a single sliding door in each side. This was mounted centrally in the side and slid to the right, the door opening being about 5ft 0in.

The vehicles with 9ft 0in wheelbase were built during the years 1924–30, three body construction variations being encountered. The first group comprised those with all-wood bodies, which were very similar to the final MR design, though strangely the latter had a longer wheelbase. The sides were formed from vertical planks about $1\frac{1}{8}$in thick, secured to the floor at their lower ends and to a wooden cantrail at the top, running the whole length of the vehicle. Tee-section uprights and flat-section diagonal braces provided additional stiffening, while the outer ends were further strengthened by angle iron corner uprights. The ends were formed from two layers of planking, the outer of which was vertical, and the inner horizontal, both layers being $\frac{5}{8}$in thick. They were also strengthened by tee-section uprights and flat-section diagonal bracing. The doors had heavy 4 × $2\frac{1}{2}$in timber frames, with a double layer of $\frac{5}{8}$in planking forming the main part of the door, and again were vertical on the outer and horizontal on the inner faces; they were recessed back from the front face of the frame. The doors ran on two small wheels fixed to brackets at the top edge, along a rail fixed to the bodyside. The lower edge of the door was restrained from moving away from the bodyside by a series of small angle brackets fixed to the body.

Three diagrams were involved in this group, the bodies of which were basically identical. The first diagram D1664 was for unventilated vehicles, see plate 10A. The second, D1676 (fig 12) had the standard arrangement of ventilation, while the last diagram in this group D1677 had twin hood ventilators at each end; roof ventilation was absent from some vehicles, while others had four ventilators arranged with three on one side of the roof. The majority of the vans in this group were unfitted when built, though a few were either piped or fully fitted. All appear to have had the normal type of Morton brake with one brake shoe to each wheel and long side levers.

Concurrently with the construction of the above vehicles there appeared, in 1924, the first of the vans with corrugated steel ends, a feature which was to become so familiar a part of the LMS and later of the BR scene. The steel ends were in two parts; the top half wrapped round the sides forming a vertical line down the side, the lower half matched the profile of the upper half where they met, but the wrapround of the sides was angled towards the centre of the van to give a longer bottom edge. Construction was basically similar to that already described, with vertically planked sides and doors, while the ends were lined with $\frac{5}{8}$in thick vertical planks. Two diagrams were involved, the first D1663 being unventilated, see plate 10B, while the second D1832 has the standard ventilation arrangement. Again the unfitted vehicles were in the majority, and all had the normal Morton brake gear when built.

The third group of vans with a 9ft 0in wheelbase featured an all steel body lined with $\frac{5}{8}$in thick vertical planking. The bodies were of steel plates rivetted together, including the roof. Two diagrams were shown for these vehicles, the first D1828 having corrugated ends of the type already described. The second diagram, D1829 was shown as having plain steel ends; only one lot was involved, built by the Gloucester Carriage & Wagon Company. As the photograph of 186000 repro-

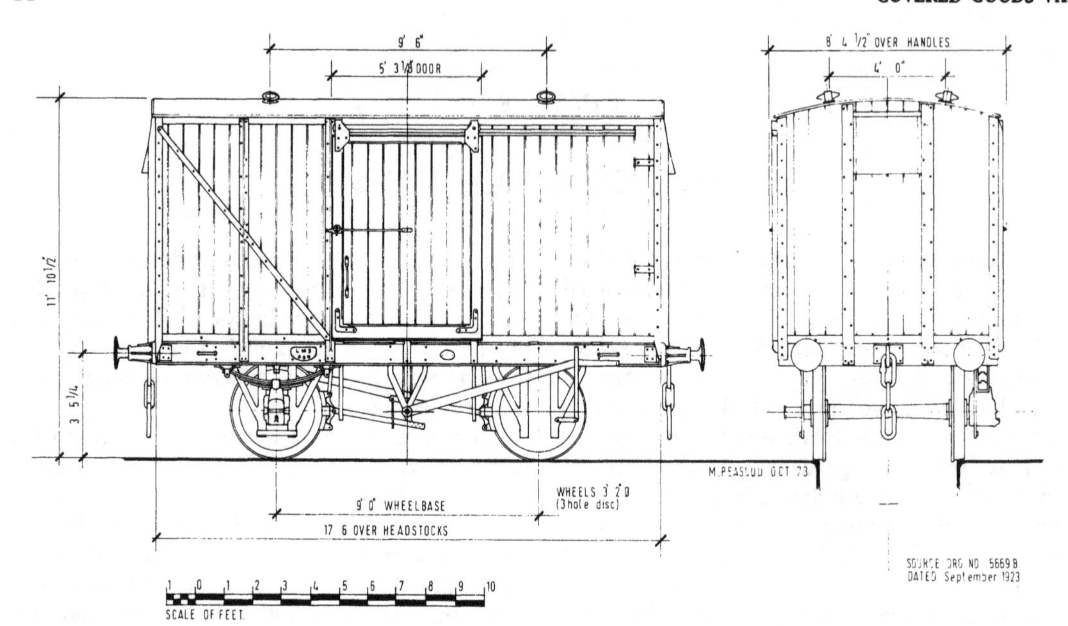

Below: Plate 10, *left upper.* Some of the first ten years' designs are illustrated in this plate commencing with 10A, showing an example of D1664 built at Wolverton in April 1924—a typical example of early livery style with neither capacity nor tare weight painted on. D1663 is illustrated in plate 10B, *right upper,* the first example of steel-end vehicles, and, like plate 10A, is an unventilated vehicle.
Plate 10C, *left lower,* is not a typical LMS van, if such a vehicle existed, for it is one of the steel vans built by the Gloucester Carriage & Wagon Company.
Plate 10D, *right lower.* is a good example of the early to mid 1930s livery style for fitted vans and shows the ex-works condition of D1814 in February 1933.

Above: Fig 12 D1676 Covered goods van

Facing page: Fig 13 D1891 Covered goods van

duced as plate 10C shows at least one vehicle of this lot with corrugated ends, the diagram book may be in error here. In both cases single hood type ventilators were fitted at each end, no roof ventilation being provided. All these vehicles were unfitted and were provided with Morton type brakes with shoes on two wheels on one side only. The corrugated steel ends obviously proved very successful, and with the introduction of the 10ft 0in wheelbase were standardised for all future construction. Nevertheless there were a good many variations on the theme.

The first vehicles had steel ends of the type already described, diagrams D1808, 1812 and 1830 showing vertically planked sides and doors. All had standard ventilation arrangement, and although little external difference was apparent the three diagrams varied in minor ways, which cannot be fully described here; they mainly concerned such dimensions as the width over door fastenings etc, while all of D1830 and some of D1808 had wheels 3ft 6½in in diameter. This group was the first to contain more fitted than unfitted vehicles; these had two brake blocks per wheel, and Morton type brakes with long handles.

The next diagram, D1814 also had the same type of ends. The sides and doors however were horizontally planked, ventilation being of the standard arrangement, see plate 10D. About a third of this type were fitted when built, and had two brake shoes to each wheel, the brake levers being of the short type. All the 10ft 0in wheelbase vehicles so far described were built in the years 1930–3, D1814 marking the change in planking style which became the standard for future construction.

Subsequent vehicles had a modified type of corrugated end. This was still in two parts, but the side wrap-round was progressively wider from top to bottom, thus forming a continuous sloping line down the side. Four diagrams were involved, namely D1891, fig 13., 1897, 1978 and 2039. There were again numerous minor variations between the diagrams, but visually the differences between the fitted and unfitted vehicles to the same diagram were more noticeable than those between the diagrams. The only certain way of distinguishing the various diagrams was by the running numbers. Planking on the sides and ends was horizontal, with prominent channel section stiffeners placed midway between the doors and ends. Ventilation in all cases was of the standard type, see plate 11A. The fitted vehicles had two brake blocks to each wheel, with short brake handles, while the unfitted variety had the normal Morton brake with long handles.

Three vehicles to D1897 were modified to produce the prototype of the well known BR pallet vans and were given diagram number 2177. The modifications entailed the provision of a door opening of 8ft 6in centrally positioned along the bodyside. To cover this large aperture, double sliding doors were provided, which slid from the centre towards each end. The date of this conversion is uncertain, but it was carried out under BR ownership, see plate 11B.

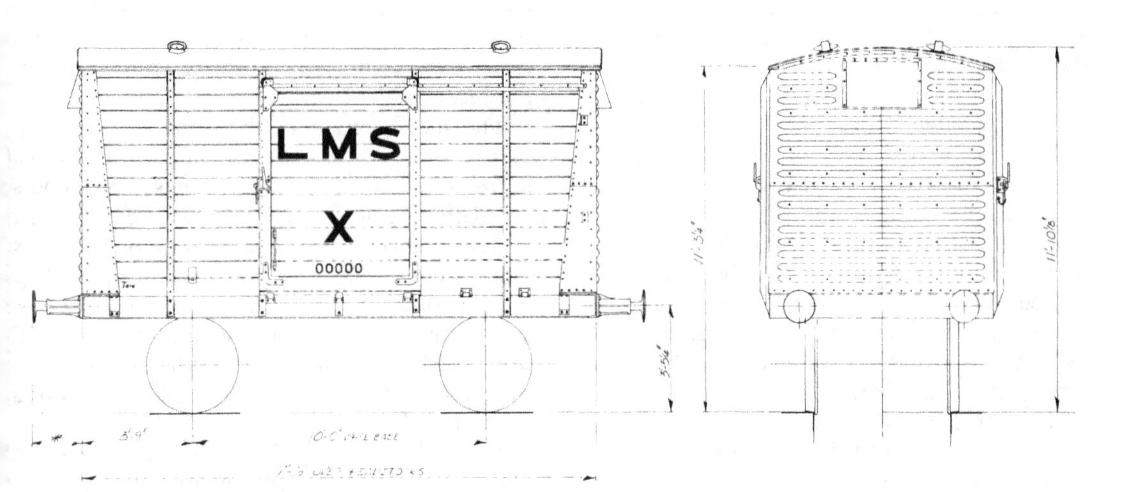

Plate 11A, *left upper,* is an example of D1978 and of the virtually similar diagrams D1891, 1897, 1978 and 2039. A special van is portrayed in as much as 511246 was branded as a fruit van to be returned to Evesham and may have had interior shelves or similar arrangements. Apart from the branding, the livery was the final style used with bauxite body colour vehicles. The British Railways modification to produce the prototype BR Pallet vans is illustrated in plate 11B, *right upper,* and allocated D2177, three vehicles being so altered.

The final style of steel-end vans is illustrated in plate 11C, *left lower,* and again is a special, branded 'FISH'. This is the only example recorded by the Authors so lettered. No 525333 is to D2108.

Shortage of steel due to war conditions was probably responsible for the reversion to wooden ends as illustrated in Plate 11D, *right lower,* an example of D2070 photographed at Earlestown in June 1964. R. J. Essery

The final vehicles in the steel ended group were those to diagrams D2097 and 2108. They were built in 1944 and had plywood sides and panelled doors. D2108 had the standard ventilation arrangement, while D2097 lacked roof ventilators. The fitted examples had two brake blocks per wheel with short brake handles, while the unfitted type had the normal Morton type brakes with long brake handles, see plate 11C.

We must now retrace our steps to 1933, to cover the 10ft 0in wheelbase vehicles of all-steel construction. Unlike the 9ft 0in wheelbase vehicles, several hundreds of which were built, this group covers five vehicles only. The diagram number was D1889, and featured corrugated ends of the early type. Externally they were similar to the 9ft 0in wheelbase vehicles to D1828, but differed in having the standard ventilation arrangement; normal Morton type brakes were fitted.

The final group of vehicles to be considered is in many ways the most interesting. As previously stated, the steel-ended type of construction was standardised from an early date. Wartime shortage of steel however led to a reversion to the all-wood type of body construction. Also due to wartime conditions, vehicles were supplied to the LMS by two of the other railway companies, either to their own design or to the modified designs produced by the LMS.

The first diagram in this group D2070, was basically of LMS design. These vehicles had horizontally planked sides, ends and doors, ventilation being of the standard type. They had diagonal bracing at the outer ends of the sides, reminiscent of Southern practice. This is perhaps not surprising in view of the fact that they were built as a part lot, supplementing vehicles with corrugated ends, all being built by the latter company; normal Morton type brakes were fitted, see plate 11D.

The next diagram, D2078, also comprised vans built by the Southern Railway, in this case of that company's standard type. Sides and ends were horizontally planked, having pairs of alternate wide and narrow planks. Double hinged doors were provided, planked in a similar manner to the sides. Twin hood type ventilators were provided at both ends, no roof ventilators being fitted. The roof itself was of the semi-elliptical type favoured by the Southern, which gave these vehicles a very distinctive appearance. Official records state that the double type of brake was fitted when built, but this appears to have been altered to the Morton type when these vehicles were vacuum fitted by BR, see plate 12A.

The LNER also provided vehicles for the LMS. They were to diagram D2079, and were basically of the building company's standard design. They had vertically planked sides and doors, while the ends were horizontally planked. The ends were stiffened by heavy wooden stanchions, and single hood type ventilators were provided; these were also of timber construction. Normal Morton type brakes were fitted, see plate 12B.

The next diagram in this group, D2088, applied to vehicles very similar to those of D2070 already described, with additional diagonal bracing on the ends; these had normal Morton type brakes and were built mainly as part lots, supplementing the normal steel ended type.

The final vehicles in this group were to D2149; they were purchased from the Ministry of Supply and were designed by the Gloucester Carriage & Wagon Company, who also built them. They were unventilated, but were otherwise very similar to normal LMS practice, normal Morton type brakes being provided.

In addition to these normal planked vehicles, two diagrams were issued for plywood panelled versions. The first, to D2103 (fig 14) were of normal LMS design, but looked somewhat plain due to the lack of planking joints. The doors, however, had additional vertical stiffeners across the panelling. Ventilation was of the standard type. Both fitted and unfitted vehicles were built, the former having two brake shoes per wheel with short brake handles; while the latter had the normal Morton arrangement. Plate 12C.

The second diagram D2112, was an experimental design for the carriage of fruit and vegetables. The body was basically the same as D2103, but with no fewer than eight roof ventilators, plus single hood vents at each end, and four scoop type ventilators along the lower edge of the bodyside. (See plate 12D). All were fitted, brake gear being as for D2103.

The last of all the covered goods to be described are those built to D2074, which were for the Northern Counties Committee. Like the open goods described in the previous chapter these were for the 5ft 3in gauge and were built by the LNER company. They were 17ft 6in long on a 10ft 0in wheelbase, and not only had all-wood bodies with heavy outside framing, but had timber underframes as well. Thus, although they were built in 1942, they displayed all of the features usually associated with the pre-grouping period. And, in fact, vehicles of practically identical appearance had been built by the Midland Railway in the early years of this century!

Livery Details

The placing of the markings on these vehicles was, like the vehicles themselves, extremely varied. It is hoped that the following paragraphs, whilst not exhaustive, will give a good idea of the main variations encountered.

The company initials were at first 12in high and were placed on the door, in the majority of cases being placed just over halfway up; in the case of the early all-steel vans, however, which had diagonal cross bracings on the door, the initials were placed at the top of the door to avoid them. At the same time some vehicles had the words VENTILATED VAN (or VENTILATED) in 5in letters (the V's being 6in) painted on the door below the initials, the spacing between them being about 12in. This latter marking appears to have been applied only to the vans with 9ft 0in wheelbase, and then not universally. During this early period the running number appeared at the left hand end of the body, and was placed about 12in above the bottom edge of the body; it was normally positioned between the vertical stiffener and the body end but cases are known where it was placed adjacent to the door. The tare weight, in all cases where it can be seen, appears to have been placed at the right hand end, at the bottom edge of the body.

The first official variation of the above scheme appeared to coincide with the introduction of the block numbering, the running number now appearing towards the bottom of the door, clear of the framing. The tare weight at this time appeared in one of four places, ie at the right or left lower edge of the body, or on the solebar, again either at the right or left hand end.

Some of the fitted vehicles at this time received the large X painted just above the running number on the door, with the non-common-user markings in the correct position at each end, although this again was by no means universal.

So far the large company initials had been used. The next scheme reduced the LMS to 4in high, though still placed on the door. Immediately below there appeared for the first time the carrying capacity expressed as 12T; at the same time the running number was moved to the left hand end of the body, with the tare weight at the right hand end. A variation of the above had the carrying capacity above the running number at the left hand end.

While it is not certain, it is thought that some vehicles in both grey and bauxite livery appeared with the above lettering scheme, which was nevertheless short-lived and soon gave way to the standard positions as recorded in chapter two.

In early LMS days, at least, vans were used and paid for mainly by the van load, not by tons of load as other traffic. The chocolate companies used to take advantage of this fact, to pack their products into the vans even into the roof arch. There was thus no need for a tare weight and this may explain its absence from early vehicles.

Fig 14 D2103 Covered goods van

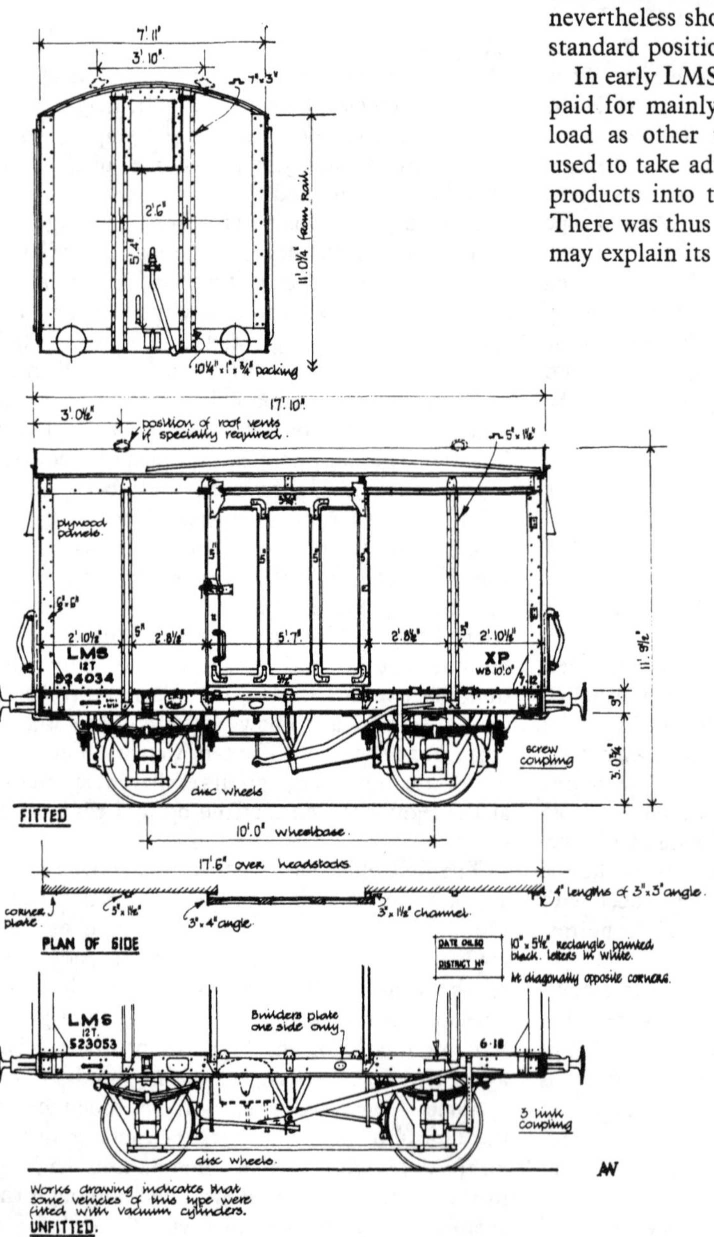

Right: Plate 12A, *left upper,* is a picture of a typical SR vehicle, built by that company for the LMS and allocated diagram No D2078.

Plate 12B, *right upper,* is a vehicle built by the LNER for the LMS, allocated diagram No D2079.

All-plywood body vehicles are illustrated in plates 12C, *left lower,* and 12D, *right lower.* The former calls for little comment while the latter is an experimental vehicle for the carriage of fruit and vegetables. Special features included improved ventilation by forced draught, double roof for insulation and four peremanently attached wire mesh shelves.

Summary of Covered Goods Vans

ALL VEHICLES HAVE STANDARD STEEL UNDERFRAMES UNLESS STATED OTHERWISE

VEHICLES WITH 9ft 0in WHEELBASE'

ALL WOOD BODY 12 ton CAPACITY

Diag	*Lot*	*Qty*	*Built*	*Year*	*Remarks*
1664	34	1000	Wolverton	1924/5	Code (unfitted) V,
	Pt36	44	Derby	1925/6	(fitted) FV, (piped PV
	107	500	Wolverton	1925/6	120 fitted 50 piped
	113	1000	"	1925/6	210 fitted 120 piped
Sample Numbers 26000, 236466, 263141, 291859, 295601, 305207, 344226					Tare 7t 4c (unfitted) 7t 6c (piped) 7t 10c (fitted)
1676	Pt36	156	Wolverton	1924/6	Code (unfitted) VV
	251	220	"	1927	(fitted) FVV
	252	110	"	1927	(piped) PVV
	253	670	"	1926/7	Tare 7t 6c (unfitted)
	337	100	Newton Heath	1927/8	7t 7c (piped)
	338	50	" "	1927/8	7t 14c (fitted)
	342	650	Wolverton	1927	
	369	334	"	1928	
	370	166	"	1928	
	387	500	"	1928	
	444	334	"	1929	

Sample Numbers 7813, 20009, 159144, 186194, 186995, 191154, 204565, 212681, 212702, 230943.

Diag	*Lot*	*Qty*	*Built*	*Year*	*Remarks*
1677	83	50	Wolverton	1925	Codes as D1676 Tare as D1664

WOOD BODY WITH CORRUGATED ENDS 12 ton CAPACITY'

Diag	*Lot*	*Qty*	*Built*	*Year*	*Remarks*
1663	73	100	Wolverton	1924/5	Codes as D1664
	186	100	B'ham C & W	1925	Fitted Tare 7t 16c
	187	100	Metro C & W	1926	Fitted
	188	150	" "	1925	
	189	100	Midland Wagon	1925	Piped Tare 7t 10c
	190	100	Hurst Nelson	1925/6	Tare (unfitted)
	191	100	S J Claye	1925	7t 6c
	192	100	Gloucester	1925	

Sample Numbers 5803, 148117, 155543, 186205

1832	443	2500	Wolverton	1929/30	Tare 7t 7c
	445	166	,,	1929	Code as D1664
	545	450	,,	1930	

Sample Numbers 150756, 156079, 156602, 157259, 160222, 197891, 203975

ALL STEEL BODY WITH CORRUGATED ENDS 12 ton CAPACITY

1828	484	400	Metro C & W	1929/30	Tare 7t 13c
	485*	300	Gloucester	1929/30	Code V
	486	150	Chas Roberts	1929/30	
	487	150	Pickering	1929/30	

This lot shown as D1829 in lot book, but see text.
Sample Numbers 178129, 183151, 184358, 184882, 186000, 187085, 197486

VEHICLES WITH 10ft 0in WHEELBASE
VERTICALLY PLANKED BODY WITH EARLY TYPE CORRUGATED ENDS 12 ton CAPACITY

Diag	*Lot*	*Qty*	*Built*	*Year*	*Remarks*
1812	544	1050	Wolverton	1930/1	Fitted Tare 8t 11c
1830	618	2000	Wolverton	1931/2	Fitted Tare 8t 11c
	619	405	Derby	1931/2	
1808	672	1100	Wolverton	1932/3	

Sample Numbers 131391, 154652, 157050, 170415, 173676, 176059, 202716, 215260, 227382, 232588.

HORIZONTALLY PLANKED BODY WITH EARLY TYPE CORRUGATED ENDS 12 ton CAPACITY

Diag	*Lot*	*Qty*	*Built*	*Year*	*Remarks*
1814	674	365	Derby	1933	Fitted Tare 8t 4c
	675	135	Wolverton	1933	
	713	249	,,	1933	
	714	250	,,	1933	

Sample Numbers 91548, 103994, 112372, 114982, 127684, 202684, 223779, 217418.

HORIZONTALLY PLANKED BODY WITH LATER TYPE CORRUGATED ENDS 12 ton CAPACITY'

Diag	*Lot*	*Qty*	*Built*	*Year*	*Numbers*	*Remarks*
1891	766	500	Derby	1934	500000–500499	Fitted
	767	499	,,	1934	500500–500998	Fitted
	768	496	Wolverton	1934	501000–501495	
	769	500	,,	1934	501500–501999	
	770	500	,,	1934	502000–502499	
	771	500	,,	1934	502500–502999	
	869	1	Derby	1934	500999	Fitted
1897	823	800	Derby	1935	506000–506799	Fitted
	824	800	,,	1935	506800–507599	Fitted
	825	400	,,	1935	507600–507999	Fitted
	826	200	,,	1935	505800–505999	
	838	1000	Wolverton	1935	503000–503999	
	839	1000	,,	1935	504000–504999	
	840	800	,,	1935	505000–505799	
	927	1000	,,	1936	508300–509299	
	928	1200	,,	1936	509300–510499	
	929	300	,,	1936	508000–508299	Fitted
2177	MODIFIED TO PALLET VANS FROM D1897. ALL FROM LOT 927. Numbers 508473, 508566, 508978.					
1978	1033	500	Wolverton	1937/8	510500–510999	Fitted
	1115	500	,,	1938	511000–511499	Fitted
	1116	500	,,	1938/9	511500–511999	
	1117	500	,,	1939	512000–512499	
2039	1112	200	Derby	1940	514125–514324	Fitted
	1272	500	Wolverton	1940	512500–512999	
	1273	500	,,	1940	513000–513499	
	1274	250	,,	1940	513500–513749	
	1275	250	,,	1940	513750–513999	
	Pt1282	396	Southern Rly	1942	514325–514720	
	1283	110	Wolverton	1940	515215–515324	
	1289	125	Derby	1940	514000–514124	
	1290	390	Wolverton	1941	515325–515714	
	1298	500	Derby	1940	515715–516214	
	1302	125	Wolverton	1941	516215–516339	
	1303	200	,,	1941	516340–516539	
	1317	800	,,	1942	516540–517339	
	1329	500	,,	1942	517340–517839	

Diag	Lot	Qty	Built	Year	Numbers	Remarks
	1330	500	Wolverton	1942	517840–518339	
	1333	800	,,	1942/3	518340—519139	
	Pt1338	500*	,,	1943	519140–519639	
	Pt1339	494*	,,	1943	519640–520133	
	1340	500	,,	1943	520140–520639	
	Pt1341	500*	,,	1943	520640–521139	
	1350	6	,,	1945	520134–520139	
	1362	500	D'by & W'ton	1944	522290–522789	

*These totals and running numbers apply to the complete lots, the balance of which were built to D2088.

PLYWOOD PANELLED BODIES WITH LATER TYPE CORRUGATED ENDS 12 ton CAPACITY

Diag	Lot	Qty	Built	Year	Numbers	Remarks
2097	1361	440	Wolverton	1944	521850–522289	
2108	1413	1350	Wolverton	1944	525140–526489	Fitted

ALL STEEL BODY WITH EARLY TYPE CORRUGATED ENDS 12 ton CAPACITY

Diag	Lot	Qty	Built	Year	Numbers	Remarks
1889	726	1	Wolverton	1933		
	792	4	,,	1934	501496–501499	

ALL WOOD BODY 12 ton CAPACITY

Diag	Lot	Qty	Built	Year	Numbers	Remarks
2070	Pt1282	494	Southern Rly	1942	514721–515214	
2078	1334	150	Southern Rly	1942	521140–521289	
	1373	250	,, ,,	1944	523290–523539	
2079	1335	250	F'dale (LNER)	1942	521290–521539	
2088	Pt1338		Wolverton	1943		for totals
	Pt1339		,,	1943		and running
	Pt1341		,,	1943		No's see D2039
	1354	850	,,	1943	521540–521849	
2149	1596	4	Gloucester	1949		

ALL WOOD BODY (PLYWOOD PANELLED) 12 ton CAPACITY

Diag	Lot	Qty	Built	Year	Numbers	Remarks
2103	1372	500	Derby	1946	522790–523289	
	1383	1594	Wolverton	1945	523540–525133	
2112	1430	6	Wolverton	1945	525134–525139	

ALL WOOD BODY AND UNDERFRAME

Diag	Lot	Qty	Built	Year	Numbers	Remarks
2074	1326	100	LNER	1942	2401–2500	For NCC

CHAPTER 6

MINERAL WAGONS

THE mineral which immediately springs to mind when talking of railways is coal, and this basically is what the vehicles described in this chapter were built to carry. Mention of rail borne coal in pre-war days also recalls the sight of long strings of private owner wagons, so beloved by enthusiasts, but with each wagon going to individual destinations led to complicated workings for the railway staff.

Little is known of the first vehicles to be described here; they were shown in the lot book only as *end door mineral wagons*, without diagram or drawing numbers. It can be taken however that they were probably very similar to the standard vehicles to be described in the next paragraphs.

In 1923 the Railway Clearing House issued a drawing to which all new mineral wagons built for private owners were to conform. The LMS immediately adopted this design for its own use; at first it was given D1343 in the MR list, but this was later amended to D1671 in the LMS series, see plate 13A and fig 15.

The main feature of mineral wagons lay in the multiplicity of doors with which they were provided, to allow the use of any of several different ways of discharging their contents. Construction was of the all-wood type, and closely followed the methods already described for the open goods. The body was formed with seven planks, and was 16ft 6in over headstocks on a 9ft 0in wheelbase. Side doors of the drop down type were provided, extending in height over the bottom five planks; the two top planks thus ran straight through the

Four vehicles, two wooden bodied, two all-steel, illustrate this chapter. Plate 13A, *left upper,* shows an example in the final livery style using large company letters and with the full length white stripe. Some variations of this exist, and a photograph shows 604889 with the white stripe ending to the right of the running number like 600425, but not carried on to the curb rail, together with 12T and LMS in the post-1936 style above the running number. The body colour of 604889 was grey.

Plate 13B, *left lower,* is the final style of wooden body vehicle built by the LMS, D2102, with steel underframe.

Below: Fig 15 D1671 Mineral wagon

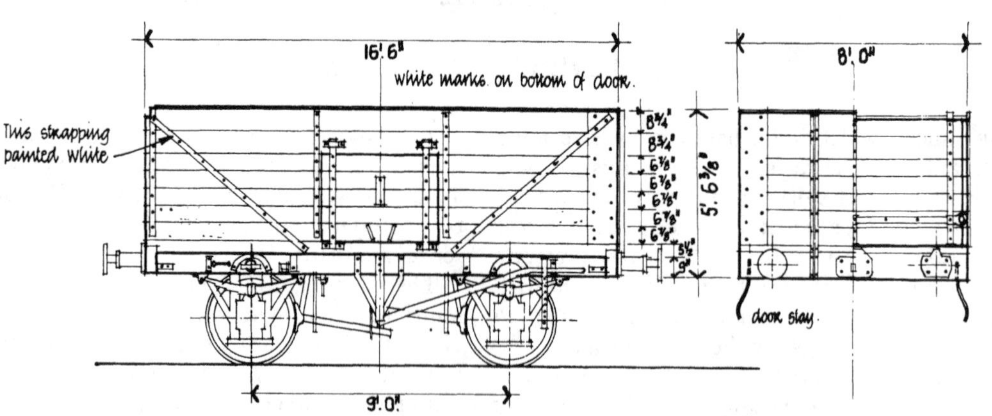

body · grey.
underframes etc. black.
lettering · white.

The steel body 16T vehicles are fully discussed in this chapter and plate 13C, *right upper,* is an example of D2109 while plate 13D, *right lower,* is one of D2134 vehicles.

length of the wagon. To allow unloading on coal drops or staithes, two bottom doors were provided, in line with the side doors, disposed equally about the centre line of the vehicle. To facilitate tipping, an end door was provided at one end, hinged across the top so that it fell open under the influence of gravity when tipped.

Unlike the vehicles described in the previous chapters, this diagram sufficed for all mineral wagons from 1924 to 1940, later lots being built to a slightly modified LMS drawing, rather than the pure RCH version, while two lots were built with steel underframes. These latter vehicles appear to have been included in D1671 at first, but were later given a separate diagram, No D2061.

After the second world war there was a general shortage of merchandise wagons, and this was partially overcome by modifying some of these vehicles. A new diagram (D2049) was issued, and this shows that the modification involved removing the portion of the top plank over the door, and fitting new or modified doors, which thus extended in height over six planks.

These vehicles were also used as the basis for rather more drastic rebuildings, those which received new diagram numbers being described in the following chapters, while one which was done, apparently without official notice is illustrated in plate 4A.

The final wooden-bodied vehicles appeared in 1944. They were to D2102 and had steel underframes, main dimensions being identical to the earlier vehicles. Thinner timber however was used for the body planking, see plate 13B. This final diagram was rated to carry 13 tons when built; most of the earlier vehicles had originally been rated at 12 tons were uprated during the war years to the higher figure.

In 1947 the LMS produced the first of its version of the 16 ton capacity all-steel wagon. This type had appeared a few years earlier, being built for the Ministry of Supply, and latterly was adopted by BR, when it became a very familiar part of the railway scene. Dimensionally the all-steel vehicles were practically identical to their wooden predecessors and had the same arrangement of doors. They featured an all-welded jig-built body, secured to the underframe as a complete unit. Four diagrams were issued, the first, D2106, being for one prototype vehicle. This was followed by D2109, these being practically identical production vehicles. (Plate 13c).

The final two diagrams differed in having small drop flap doors above the main side doors; D2134-plate 13D were shown as 16 ton mineral wagons while D2153 were described as 27 ton iron ore tipplers. This however may be an incorrect description, as most tipplers lacked doors.

On all of these vehicles there were considerable variations within each diagram. Some vehicles had pressed steel doors, both end and side, while others had pressed steel doors either on the end or sides, the other door(s) being made from steel plate with welded stiffeners. Again others had all doors of the plate type. This was probably due to steel shortages which prevailed at the time, and the availability of pressings; vehicles urgently required for service were probably fitted with plate doors to save time, see plates 13C and 13D. All the vehicles in this chapter were unfitted when built, and all had the double type of hand brake.

Livery Details

The all-wood vehicles to D1671 at first had 18in high company initials. They were placed either on the three middle planks, or were displaced one plank lower; in the former position they were clear of the strapping, while in the latter position both the L and S ran partly over the diagonal straps. Vehicle number was on the bottom plank at the left hand end, with the tare weight on the curb rail below and slightly to the right. At a somewhat later date the white diagonal stripe denoting the end door was added, the date of introduction of this feature is not certain but it was in use by 1928. The stripe ran from the top of the wagon, at the door end, to the bottom plank at the opposite end, and was broken where it crossed the company initials. On the later lots the stripe ran on to the curb rail. In all cases it was about 4in wide.

The last lots of D1671 came out in bauxite livery, and these had the insignia in the correct positions. They retained the long diagonal stripe, and in addition had the bottom door markings in the form of two short white stripes painted on the door/curb rail in a vee-formation. On some of these vehicles the end door marking was modified to the new standard where the stripe was painted down the diagonal strapping at the door end only.

The steel-underframed vehicles to D2102 came out with the woodwork unpainted, apart from patches on the bottom plank at each end and on the side doors. Carrying capacity, company initials and running number were painted on the left hand patch in that order, with the tare weight at the right hand end. Bottom door and end door markings were applied as previously described, the end door marking having a narrow bauxite border painted on the timber to make it more visible. The all-steel vehicles built by the LMS came out in bauxite livery; apart from this the insignia style was as outlined above for D2102, with the carrying capacity, company initials and running number painted in line rather than above one another in the officially correct manner.

Summary of Mineral Wagons

END DOOR MINERAL WAGON NO DIAGRAM

Lot 159 Qty 1500 Built Newton Heath Year 1926

ALL-WOOD TYPE 12 ton CAPACITY D1671
CODE EDW TARE 7 tons CUBIC CAPACITY 534 cu ft. Double Brake

Lot	Qty	Built	Year
40	250	Trade	1924
41	250	,,	1924
42	150	,,	1924
43	150	,,	1924
44	250	,,	1924
45	400	,,	1924
46	250	,,	1924
47	250	,,	1924
48	350	,,	1924
49	250	,,	1924
50	250	,,	1924
51	175	,,	1924
52	500	,,	1924
53	250	,,	1924
54	100	,,	1924

Lot	Qty	Built	Year
55	225	Trade	1924
59	1	Derby	1924
108	1000	Earlstown	1924/5
118	2000	,,	1925/6
161	1650	,,	1926
162	150	Butterley	1925
163	100	Chorley Wag	1925
164	200	S J Claye	1925
165	150	Clayton	1925
166	150	Hall Lewis	1925
167	100	Harrison Camm	1925
168	100	T Hunter	1925
169	250	Metro C & W	1925
170	250	Midland Wag	1925
171	150	Pickering	1925

Lot	*Qty*	*Built*	*Year*
172	250	C Roberts	1925
254		Trade	1927
255		,,	1926
256		,,	1926
257	2000 for	,,	1926
258	lots 254–	,,	1926
259	267	,,	1926
260		,,	1926
261		,,	1926
262		,,	1926
263		,,	1926/7
264		,,	1926
265		,,	1926/7
266		,,	1926
267		,,	1926/7
294	1400	Earlstown	1926/7
324	1000	,,	1927
330	1000	,,	1926/7
347	1500	,,	1928

Lot	*Qty*	*Built*	*Year*
380	1000	Earlstown	1928
383	100	,,	1928
386	1000	,,	1928
422	700	,,	1928
494	2000	,,	1930

Lot	*Qty*	*Built*	*Year*	*Numbers*
830	1000	Derby	1935	600000–600999
831	1000	,,	1935	601000–601999
832	500	,,	1935	602000–602499
915	500	,,	1936	602500–602999
916	1000	,,	1936	603000–603999
917	1000	,,	1936	604000–604999
1006	500	,,	1937	605000–605499
1021	1000	,,	1937	605500–606499
1022	500	,,	1937	606500–607999
1108	500	,,	1938	608000–608499
1109	500	,,	1938	608500–608999
1206	546	,,	1940	609000–609545

Sample Numbers 108203, 171253, 179653, 256180, 268985, 315512, 351270

MERCHANDISE WAGONS CONVERTED FROM D1671 D2049

100 VEHICLES CONVERTED

Sample Number 85299

WOOD BODY STEEL UNDERFRAME 12 ton CAPACITY

Diag	*Lot*	*Qty*	*Built*	*Year*	*Numbers*	*Remarks*
2061	384	100	Earlstown	1928	Various	
	385	100	,,	1928		
2102	1368	1500	Wolverton	1944/5	609546–611045	Tare
	1369	1000	,,	1944/5	611046–612045	7t 6c
	1370	500	,,	1944/5	612046–612545	
	1379	500	Derby	1945	612546–613045	
	1386	2000	Wolverton	1946	613046–615045	

ALL STEEL TYPE 16 ton CAPACITY

Diag	*Lot*	*Qty*	*Built*	*Year*	*Numbers*	*Remarks*
2106	1412	1	Derby	1947	616000	
2109	1415	1999	Derby	1947	616001–617999	
	1468	600	,,	1947	618000–618599	
2134	1516	1900	Derby	1949	618600–620499	
	1540	1600	,,		620500–622099	
2153	1609	400		1949	622100–622499	

CHAPTER 7

OPEN WAGONS FOR SPECIALISED TRAFFIC

In the previous chapters we have dealt with those vehicles which may fairly be described as having been the basic requirement for goods service in the LMS era. The brake van was, as we have seen, an essential part of every goods train, while the open and covered goods wagons, by sheer weight of numbers, were almost as essential. Mineral wagons are the first example of vehicles built to carry one particular traffic, but were so widespread as to be universal. In other words, the vehicles so far described could have been seen by the casual observer in even the smallest goods yard, or at any line-side vantage point.

The vehicles to be described in this and subsequent chapters were of a more specialised nature, and while some were quite widespread, others spent all of their working lives on one service; depending on the location of the observer they might have been seen frequently or not at all. These vehicles however still come into the ordinary wagon book, and must not be confused with those vehicles which come into the *Special Wagon* diagram book, and which are in an entirely different category.

Right: Fig 16 D1888 Sand wagon

The diverse nature of traffic and space limitations have combined to limit the many different types of vehicles in this chapter to illustrations of one or two examples of each type.

Plate 14A, *top left,* is an example of a sand wagon and while no illustrations of them in bauxite livery are known to exist, it is thought they generally followed the style of open merchandise wagons.

Plates 14B and 14C, *centre left,* show the first, D1675 and later, the longer D1945 tube wagons and also serves to illustrate the two livery styles described in the chapter. A. E. West (14B)

Plates 14D, 15A and 15B (see page 60) illustrate some of the LMS designs of long low wagons, later titled Plate Wagons by the LMS and adopted as a description by British Railways. No. 43330, plate 14D, *bottom left,* is the first livery style.

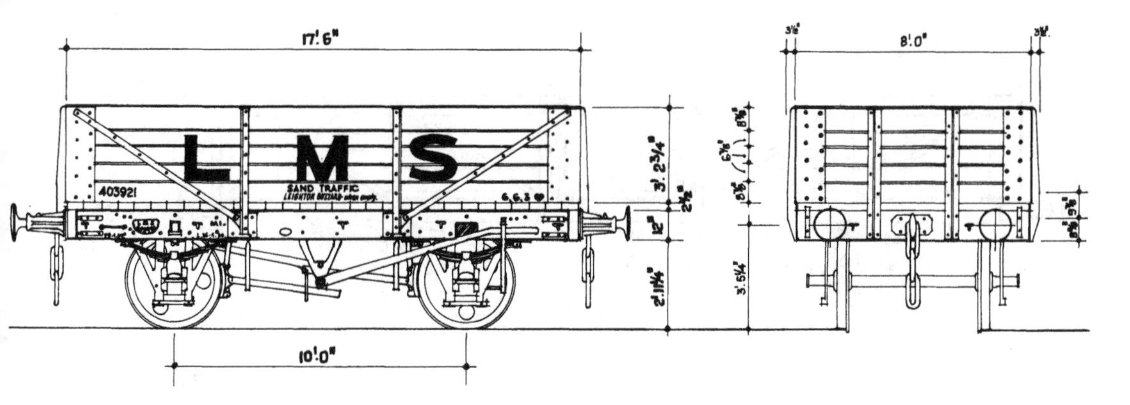

Sand Wagons

ONLY one diagram was involved here. At first glance the vehicles to D1888 could easily be taken for high-sided goods wagons. Construction was of the all-wood type, on a standard underframe with a 10ft 0in wheelbase. The body was constructed from five planks, without curb rails or doors of any type, the gaps between the planks being sealed with pitch to prevent the contents from seeping out. They were unfitted, hand brakes of the standard Morton type being fitted, see plate 14A and fig 16. These vehicles were for traffic sand, that is sand for building purposes as opposed to loco sand, the vehicles for which were, in the main, converted from open goods wagons fitted with a sloping roof.

Livery Details

Livery appears to have followed the open goods style, the only point to note being that the wagons were branded to be returned to a specific point, in most cases as in plate 14A, Leighton Buzzard.

Summary of Sand Wagons

Lot 842 Qty 100 Built Derby 1934
Numbers 403900–403999

Tube Wagons

THESE vehicles were used for the conveyance of long light loads, as well as the tubes from which they took their title. In appearance and construction they resembled a long high-sided goods wagon, having five plank bodies with a full height door in each side. All had steel underframes of 10in deep section, wheelbase in all cases being 17ft 6in.

Three main diagrams are involved, the first, D1675, comprising wagons built between 1925–36. They had a carrying capacity of 20 tons and were 27ft 0in over headstocks, brakes of the normal Morton type being fitted, see plate 14B and fig 17.

The next diagram, D1945, applied to wagons built between 1936–42. They were also of 20 tons capacity, and were very similar to the previous type. Length over headstocks was increased to 30ft 6in, while the brakes were of the 20 ton standard type, see plate 14C.

The final diagram in this group, D2116, relates to wagons built in 1947. They had the same dimensions and brake gear as D1945, but differed in that they were rated to carry 22 tons and had wood lined corrugated steel ends.

Beside the main types described above there were two odd vehicles which also come into this category. The first was very similar to D1675, and was purchased from the disposal board in 1927. This body was formed to dispose of vehicles built for the government during world war 1. The second vehicle, allocated D1987, was converted from a crane match truck in 1937. It was a bogie vehicle, rated to carry 30 tons, and was 36ft 0in over headstocks. The bogies had a 5ft 6in wheelbase, and were set at 25ft 0in centres. The body was of all-wood construction, formed from two planks, the bodysides which appear to have been of the fixed type being without doors; a handbrake of the screw down type was provided.

Livery Details

The first livery of these vehicles was again very similar to that employed on the high-sided goods. The word Tube and non-common-user markings appear to have been applied to the first vehicles with block numbering, this being modified slightly on the vehicles to D1945 to read Long Tube.

Unfortunately we have been unable to trace any photographs of D1675 in bauxite livery. However plate 14C illustrates D1945 in this livery. The vehicles to D2116 came out with the woodwork unpainted, except for the two lower planks at the left hand end painted as far as the first diagonal strap with carrying capacity, company initials and

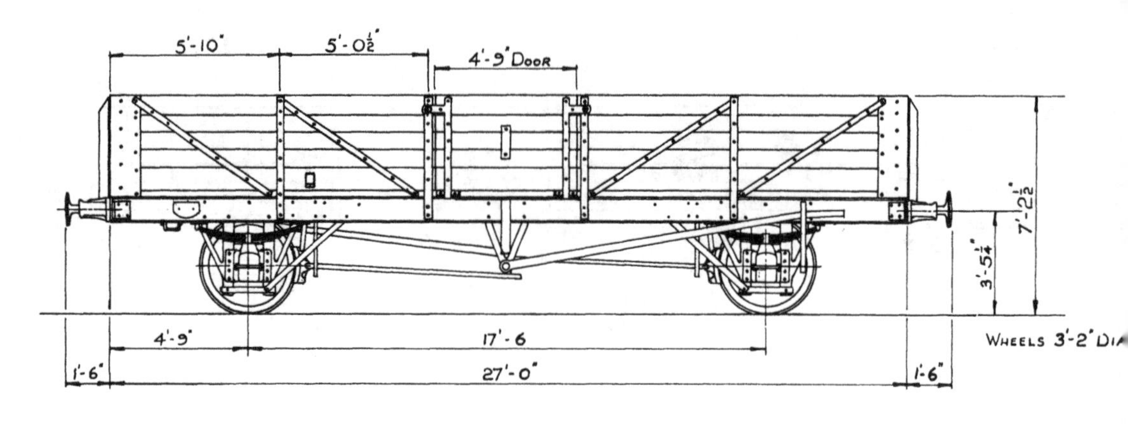

Summary of Tube Wagons

Diag	Lot	Qty	Built	Year	Numbers	Remarks
1675	143	100	Trade	1925		Tare
	144	50	"	1925		9t 5c
	145	100	"	1925	Various	Code TUW
	673	100	Wolverton	1933		
	758	100	"	1934		
	837	100	"	1935	499000–499099	
	938	200	Chas Roberts	1936	499100–499299	
	965	200	" "	1936	499400–499599	

Sample Numbers 6228, 10941, 13269, 341874.

Diag	Lot	Qty	Built	Year	Numbers	Remarks
1945	939	100	Hurst Nelson	1936	499300–499399	Tare
	1032	400	Wolverton	1937	499600–499999	9t 16c
	1142	300	"	1938/9	492000–492299	
	1320	100	"	1942	492300–492399	Code TUW
2116	1454	300	Derby	1947	492400–492699	
1987		1		1937	28495	
	314	1	Derby	1927	153964	ex disposal board

number painted on the lower plank in that order, with TUBE on the plank above. A similar patch at the right hand end on the second and third planks from the bottom carried routing instructions, an example being: 'Empty to Corby and Weldon LMS' (492464), painted in block capitals about 3in high.

Long Low Wagons

THESE were also long-wheelbase vehicles, their later title of plate wagon accurately describing the type of load they carried. The bodysides were very shallow, in general being about 12in deep. Underframes were of the steel type, and unless otherwise stated were built from 10in deep sections.

The first were built in 1927 to D1680, with a carrying capacity of 20 tons, 25ft 0in long over headstocks on a 12ft 3in wheelbase. The body was of all-wood construction, formed from two planks each 6¾in deep, the sides being of the fixed type. They were unfitted, handbrakes of the double type being provided, see plate 14D.

The next diagram D1798, was for vehicles of very similar appearance, but considerably longer. They were built in 1930, and were again of 20 ton capacity, length over headstocks being 31ft 0in on a 17ft 6in wheelbase. The body was of the all-wood type, with fixed sides formed from one plank 12in wide. They were unfitted and were provided with a handbrake of the Morton type (fig 18).

The next two diagrams had sides which could be dropped down in two sections. Carrying capacity was again 20 tons. The first diagram D1924 was for wagons built between 1933–40 having the same main dimensions as D1798. The body was of the all-wood type, and had the sides formed from a single-plank. Replacement sides however were often made from two planks, in some cases only half of the side being so treated. These were also unfitted, but although the diagram states that they were to have Morton type brakes, photographs show that some had the so-called 20 ton standard type as described for the tube wagons.

The second diagram, D2069, was for wagons built in 1942 and while retaining the 17ft 6in wheelbase they were somewhat shorter at only 27ft 0in over headstocks. Apart from this they

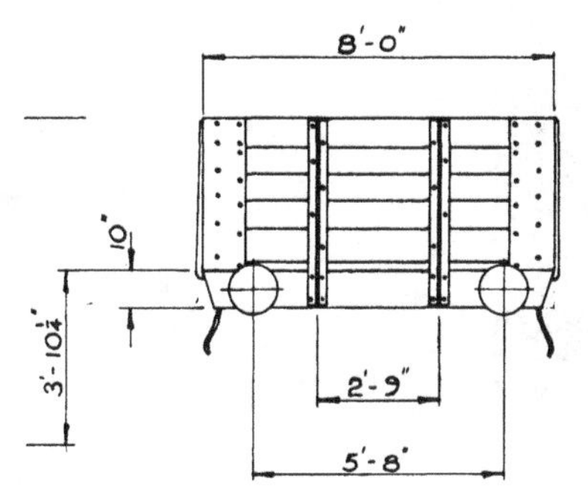

Left: Fig 17 D1675 Tube wagon

Below: Fig 18 D1798 Long low wagon

were practically identical to D1924, all having the 20 ton type of brake, see plate 15A.

The final diagram D2083 relates to the all-steel type, built between 1944–9. These vehicles were rated to carry 22 tons, the underframe built from 12in deep sections being 27ft 1½in over headstocks on a 15ft 0in wheelbase. The body was built from steel plate, reinforced with angle iron and both welded and rivetted construction appears to have been employed. The sides were again arranged to drop down in two sections, while the brake gear was of the 20 ton standard type, see plate 15B.

Livery Details

This class of vehicle is notable for a change of designation during the period of LMS ownership, starting as Long Low and becoming Plate C1944. The original livery adopted for D1680 was for LMS to be spaced out covering the two planks on a grey body. The running number was placed at the left hand end, together with details of tare and load, ie '20 tons evenly distributed.' D1798 was similar except that the LMS occupied the full depth of the side, these vehicles only being one plank high. Plate 14D can be considered typical and the tare weight 9.10.2 can just be seen on the original print, to the left of the first piece of strapping, painted on the edge of the wagon floor. Other vehicles built during this period had the tare and tonnage in the space between the corner ironwork and first vertical strapping with the running number to the right of the strapping, while others had the tare in the same place as 43330, but with the positions of the tonnage and running number reversed.

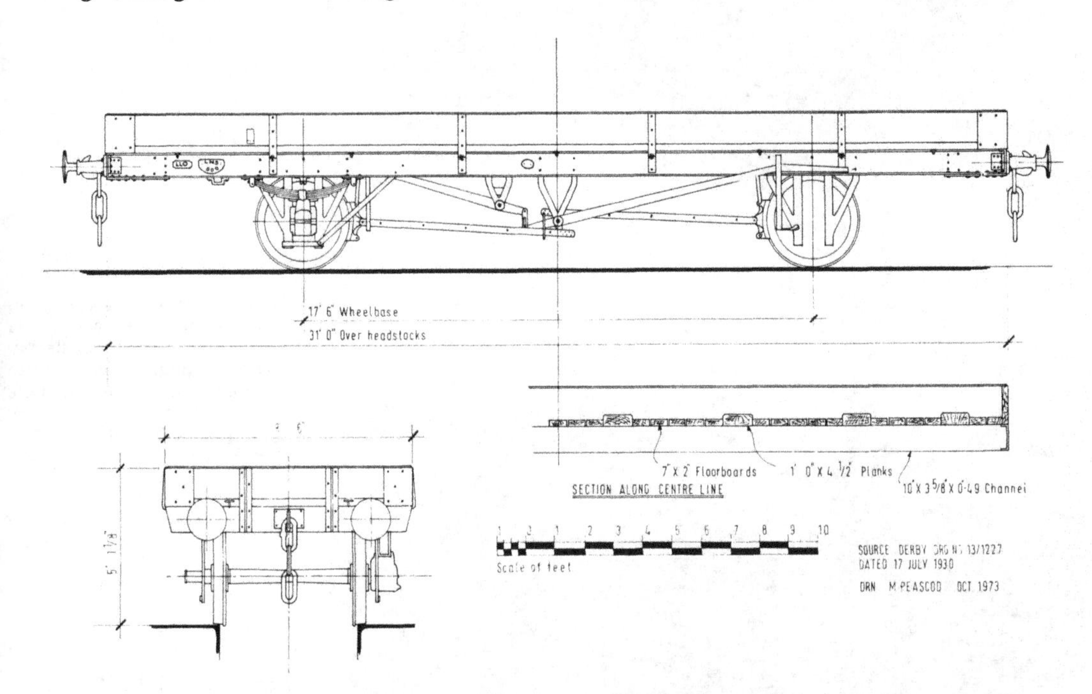

Plate 15B, *centre upper,* shows 498866 in the final LMS style for plate wagons with bauxite body colour, while M498545., plate 15A, *top,* is in typical British Railways, condition, photographed in 1964. R. J. Essery

Plate 15C, *centre lower,* is an example of a service vehicle D2098 and shows a sleeper wagon. Compare with plate 15D, *bottom,* D1954, and note that virtually the only difference is in the length of body and design of ends.

Right: Fig 19 D1954 Ballast wagon

Summary of Long Low Wagons

Diag	Lot	Qty	Built	Year	Numbers	Remarks
1680	327	187	Derby	1927		Tare
	Sample Numbers 292523, 153930					7t 13c 2q
1798	455	200	Derby	1930		Code LLO
	533	200	Newton Heath	1930		Tare 9t 9c
	Sample Numbers 43160, 43330, 43720, 45461, 46346, 324599, 352649.					
1924	836	250	Wolverton	1933	497000–497249	Tare
	926	325	,,	1936	497250–497574	9t 9c
	968	200	,,	1936	497575–497774	
	1005	250	,,	1937	497775–498024	
	1143	200	,,	1939	498025–498224	
	1215	100	,,	1940	498225–498324	
2069	1321	300	Wolverton	1942	498325–498624	Tare 9t 10c
2083	1344	250	D'by & W'ton	1944	498625–498874	Tare 9t 6c
	1355	100	Wolverton	1944	496000–496049	50 for LNER
	1367	50	,,	1944	496050–496099	
	1385	100	Derby	1946	496100–496199	
	1514	50	,,	1949	496200–496249	

With the change to bauxite body colour, the variations were reduced and the livery layout of plate 15B can be considered typical, except that vehicles built to D1924 from 497736 upwards had the tonnage above the running number at the extreme left hand end of the vehicle with the company ownership to the right of the tonnage. The tare weight was at the extreme right hand end of the vehicle.

In all livery styles, N (not in common use) was at the two extreme ends of the vehicles, when designated Long Low but not when designated Plate.

Ballast Wagons

These vehicles belonged to the Civil Engineer's Department, and they were a familiar sight to the Sunday traveller, who would find himself travelling slowly past a long line of them, attended by a large gang of permanent way men.

Two diagrams were issued for these vehicles, which as they were practically identical can be considered together. The vehicles had a carrying capacity of 12 tons, and were built on steel underframes 20ft 8in over headstocks on a 10ft 0in wheelbase; they were unfitted and had normal Morton type handbrakes. Bodies were of the all-

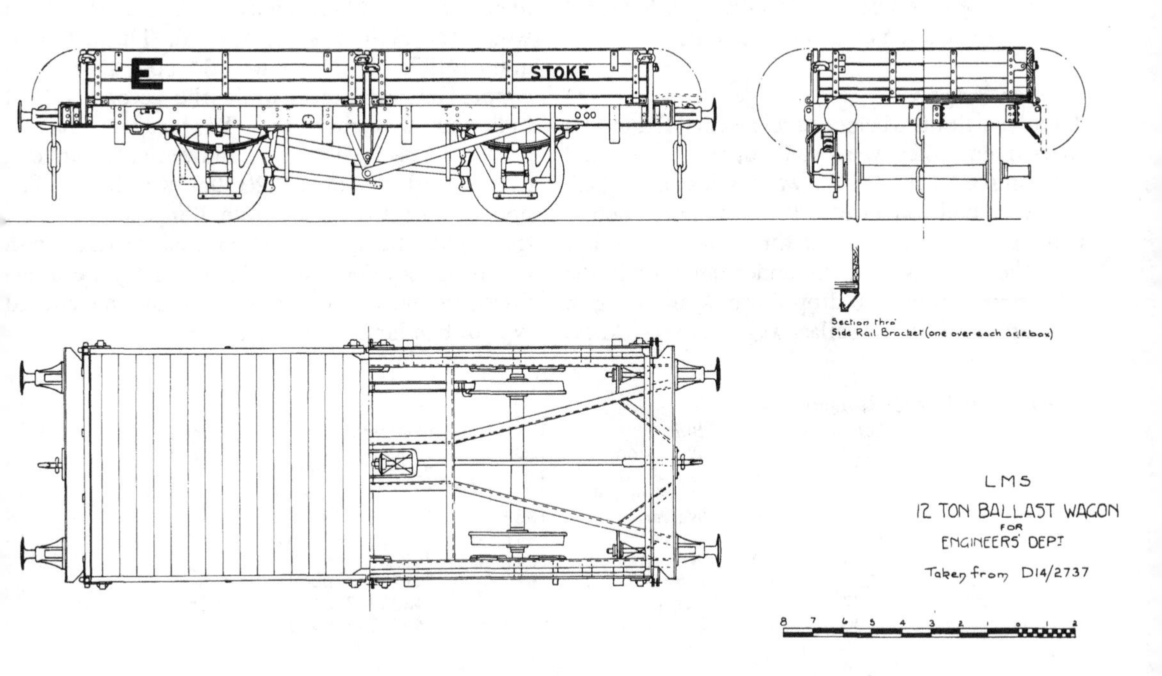

Summary of Ballast Wagons

Diag	*Lot*	*Qty*	*Built*	*Year*	*Numbers*	*Remarks*
1954	978	60	Wolverton	1936	740000–740059	Tare
	1029	500	Derby	1937	740060–740559	6t 8c
	1106	500	"	1938	740560–741059	
	1107	500	"	1938	741060–741559	
	1209	350	"	1941	741560–741909	
	1218	350	Wolverton	1939/40	741910–742259	
	1285	250	"	1941	742260–742509	
	1318	250	D'by & W'ton	1942	742510–742759	
2095	1347	100	D'by & W'ton	1944	742760–742859	Tare
	1389	100	Derby	1946	742860–742959	6t 10c
	1426	100	"	1946	742960–743059	
	1479	100	D'by & W'ton	1947	743060–743159	

wood type, and were formed from three planks. The sides and ends were arranged to drop down, the sides being in two portions. The body was somewhat shorter than the underframe, so that the ends when dropped rested on the top of the underframe, and in this condition the vehicles could be used for the carriage of rails.

Construction period for the two diagrams overlapped, D1954 (fig 19) being built between 1936–49, while D2095 was built between 1944–7. The main difference between the two types lay in the thickness of timber used for the bodies and although both types were 19ft 0in long inside, outside dimensions were 19ft 4¾in and 19ft 2¾in respectively. Similarly although the width outside was 7ft 6in in both cases the inside dimensions were 7ft 1¼in and 7ft 3¼in, see plate 15C.

Livery Details
Plate 15C illustrates the only known photograph of one of these vehicles in LMS livery with what is believed to be a red oxide body colour.

Sleeper Wagons
THESE vehicles also belonged to the Civil Engineer's Department; they were contemporary with and very similar to the ballast wagons just described, and were built on identical underframes. Bodies were again of the all-wood three plank type, but were the full length of the underframe. Only the sides were arranged to drop down, again being in two portions. Like the ballast wagons, the thickness of the body timbers was the main difference between the two diagrams; both were 20ft 3¼in long inside, D1953 being 7ft 1¼in wide inside while D2098 was 7ft 3¼in, see plate 15D.

Livery Details
The original photographs suggest a light grey body colour, while observers claim that red oxide which weathered pink was the body colour adopted. Certainly wartime construction was unpainted and the matter is discussed in chapter 2.

Locomotive Coal Wagons
THE purpose for which these vehicles were built is obvious from their title, and they were rarely if ever used for any other traffic. Unlike the ordinary mineral wagons, they worked only to railway depots, and were not so restricted to size by the same considerations which applied to the former.

The LMS issued three diagrams for these vehicles, of which the first two were externally identical, and differed only in the material from which the body was constructed. The first diagram D1973 (fig 20) showed the body built from copper bearing steel, while the second D1974 indicated wrought iron. The wagons had a carrying capacity of 20 tons and were 21ft 6in long over headstocks on a 12ft 0in wheelbase. The underframes were built from 10in deep sections, the bodies being of rivetted construction, with two drop flap doors in each side. They were unfitted, and were provided with the 20 ton standard type of handbrake, see plate 16A.

Summary of Sleeper Wagons

Diag	*Lot*	*Qty*	*Built*	*Year*	*Numbers*	*Remarks*
1953	979	20	Wolverton	1936	749000–749019	Tare
	1078	14	"	1937	749020–749033	
	1163	75	Met Cammell	1938	749034–749108	
	1217	50	Wolverton	1939	749109–749158	
	1270	50	Derby	1941	749159–749208	
	1284	50	Wolverton	1941	749209–749258	
2098	1365	100	D'by & W'ton	1944	749259–479358	Tare
	1427	50	Derby	1946	749359–749408	6t 9c
	1480	50	D'by & W'ton	1947	749409–749458	

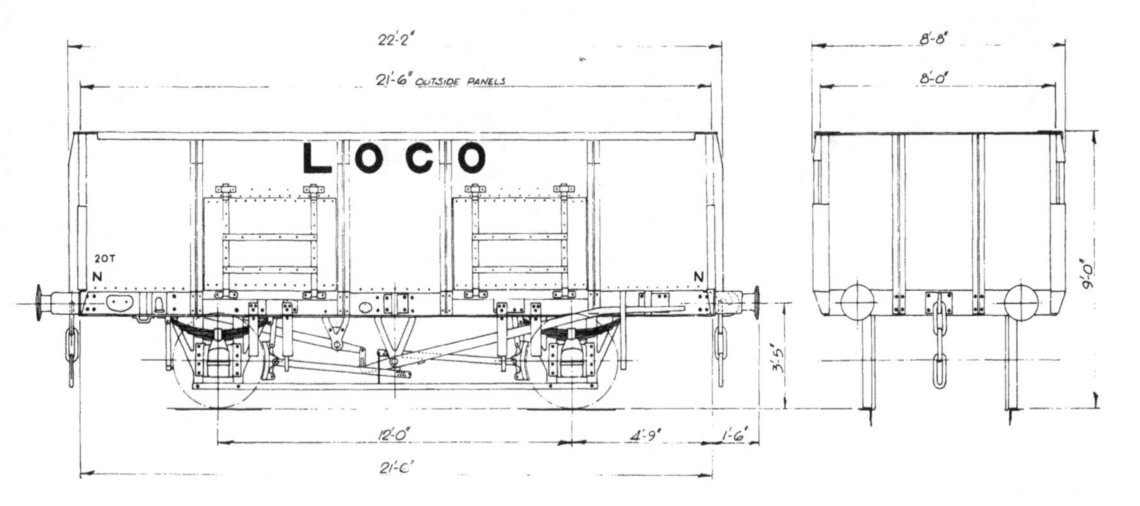

Fig 20 D1973 Loco coal wagon

Summary of Locomotive Coal Wagons

Diag	Lot	Qty	Built	Year	Numbers	Remarks
1973	990	100	G R Turner	1936	750000–750099	Tare
	991	250	Met Cammell	1936/7	750100–750349	10t 3c
	992	150	B'ham C & W	1936/7	750350–750499	Code LW
1974	993	300	Gloucester	1936/7	750500–750799	
	994	200	Fairfield	1936/7	750800–750999	
	1366	150	Derby	1945	751000–751149	
	1388	100	,,	1946	751150–751249	
	1425	100	,,	1946/7	751250–751349	
2038	1207	104	Derby	1940	757000–757103	Tare 6t 14c
	1265	150	,,	1940	757104–757253	

The final diagram was for 13 ton capacity vehicles of the all-wood type. These wagons to D2038 were very similar to the ordinary mineral wagons, and were 16ft 6in long on a 9ft 0in wheelbase. The body was of the seven plank type, and a single drop flap door was provided in each side; they were unfitted, the handbrake being of the Morton type.

Livery Details

The first lots of the 20 ton wagons appear to have been turned out in grey livery, but with the later standard lettering style. The word LOCO was painted centrally on the bodyside, just below the top edge, and was about 12in high. Non-common-user markings were applied in the correct positions, with carrying capacity and company initials above the left hand one; the running number was disposed slightly to the right. With the change to bauxite livery all of these insignia were placed in one line along the lower edge of the body at the left, the word LOCO being placed just above the company initials in 4 or 5in letters. Unfortunately we have been unable to trace details of the wooden vehicles, but it is possible that they followed this latter scheme.

Single Bolsters

THESE vehicles were the smallest built by the LMS, and were used mainly for the carriage of timber as felled, or as runners between double and bogie bolsters and adjacent vehicles where loads overhung the ends of the longer vehicles.

Two diagrams were issued, the first D1950 (fig 21) being for the all-wood type of construction, while the second diagram D2041 differed only in specifying a steel underframe. Both types were 15ft 6in long over headstocks on a 8ft 0in wheelbase, and were rated to carry 12 tons. The bodies were of the single-plank fixed-sided type, the single transverse bolster being placed centrally on the body, being arranged to have a limited swivelling movement. They were unfitted, and were provided with a Morton type handbrake, see plate 16B and fig 21.

Livery Details

Part of the first lot to D1950 appear to have been painted grey when built. These wagons had non-common-user markings on the corner plates. The company initials were placed centrally and were about 7in high; to the right the word SINGLE was painted in letters about 6in high. Running

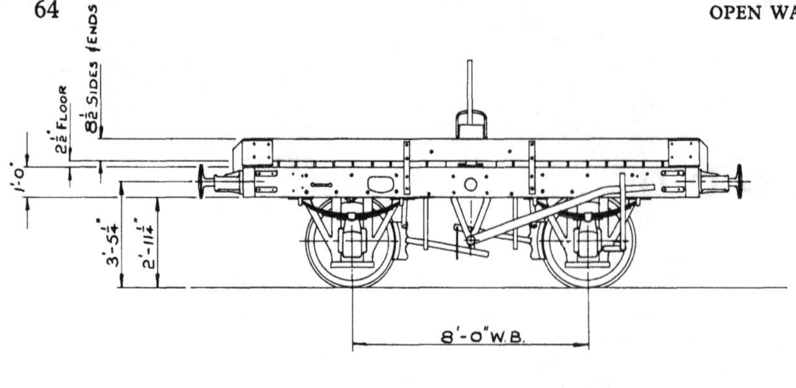

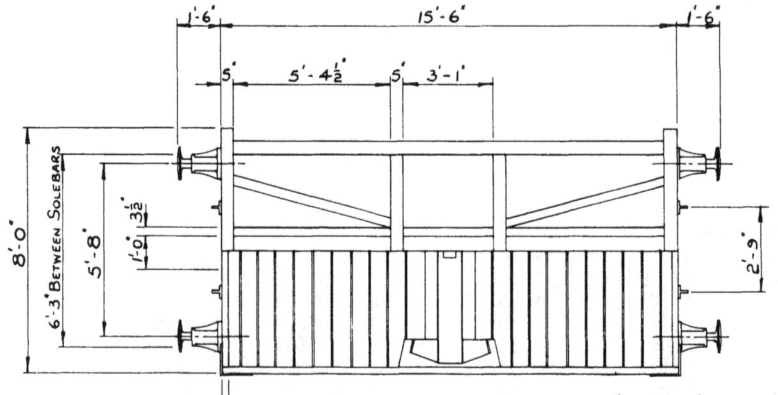

Fig 21 D1950 Single bolster truck

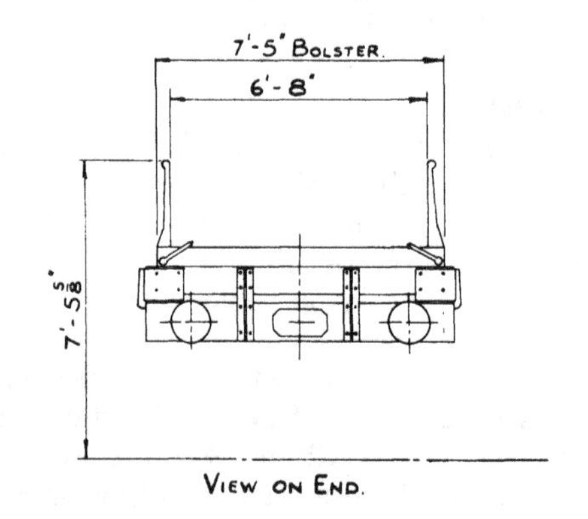

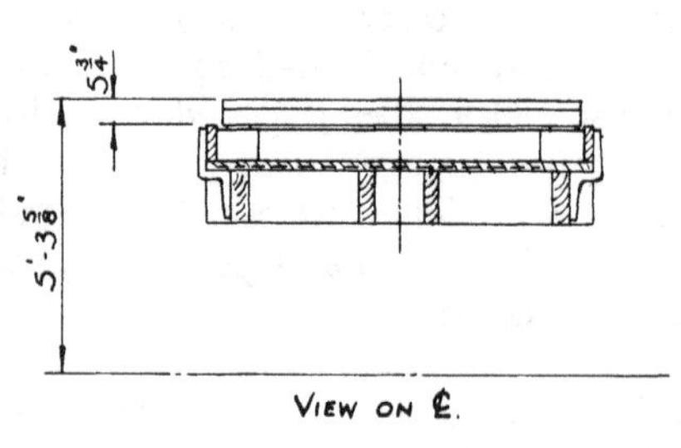

number and carrying capacity were at the left hand end, with tare at the right hand end.

With the change to bauxite livery, the company initials were moved towards the left hand end, the word SINGLE being kept in approximately the same position, but reduced like the initials to 4in in height, while the non-common-user markings disappeared altogether.

Double Bolsters

THESE were long four-wheeled vehicles, and as their title suggests were fitted with two bolsters. Like the single bolsters they were used for the carriage of timber and similar long loads, and were also used as runners for other vehicles.

Dimensionally, they fall into three groups of which the first were 27ft 0in over headstocks on a 17ft 6in wheelbase. Carrying capacity was 20 tons, and all had steel underframes made from 10in deep sections. Those of the first and second diagrams, D1674, built between 1925–30, and D1949 built between 1936–8, were practically identical

Summary of Single Bolster Wagons

Diag	*Lot*	*Qty*	*Built*	*Year*	*Numbers*	*Remarks*
1950	977	500	Derby	1936	722000–722499	Code SBK
	1105	500	,,	1938	722500–722999	Tare 6t 6c
	1201	250	,,	1939	723000–723249	
2041	1280	162	Derby	1940	723250–723411	Code SB Tare 5t 19c

Plate 16A, *top right,* is an example of a steel-bodied vehicle to D1973, built at a time when the livery style was changing. Body colour is still grey but lettering and numbers are in the new size and position.

Plate 16B, *centre upper,* shows a single bolster wagon in service, photographed on 27 August 1938; it is an example of this type of vehicle built and painted in the then new style with bauxite body colour. A. E. West

Two examples of double bolster wagons are shown in plates 16C *centre lower,* D1674, and plate 16D, *bottom,* also D1674 but altered to a canopy wagon. Unlike those converted to trestle plate wagons from D2105, whose numbers are known the BR conversion to canopy wagons cannot be traced and it is not known how many were converted. The photograph of 317842 clearly shows one of their uses in traffic, although sometimes over-hanging loads required adjoining under-runners–low-sided goods wagons–to be used with them. The rules required that when runners were used they should be behind the main vehicle, not preceding it. A. E. West (16C)

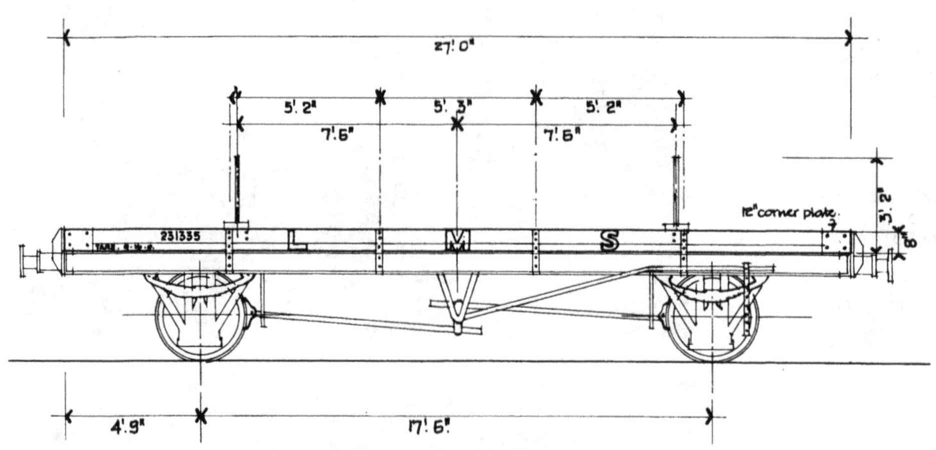

and had all-timber bodies with low fixed sides. (Plate 16C and fig 22). The only points of difference between the two were that the earlier vehicles had lower bolsters and Morton type hand brakes, the later vehicles having the 20 ton standard type of brake. Photographs show that some of the earlier vehicles had the later high type of bolster and brake, and the only certain way of distinguishing the two was by the running numbers.

The third diagram D2067, had sides 12in high arranged to drop down in two sections, and were practically identical to the long low wagons to D2069, apart from the bolsters. As the diagram stated that these could be removed it becomes obvious that these vehicles could be used as long low wagons when occasion demanded. Hand brakes of the 20 ton standard type were provided. The bolsters on the first two diagrams were at 15ft 0in centres, while those on D2067 were at 15ft 6in centres.

The other two groups have only one diagram in each, of which the first (D2029), were built in 1927. They were 25ft 0in over headstocks on a 12ft 3in wheelbase. They were fitted with hand brakes of the double type, but were otherwise very similar in appearance to the vehicles to D1674.

Summary of Double Bolsters Wagons

Diag	*Lot*	*Qty*	*Built*	*Year*	*Numbers*	*Remarks*
1674	209	75	Met Cammell	1925/6		Tare 8t 15c
	210	75	B'ham C & W	1925		Code DBQ
	211	50	Clayton	1925		
	277	100	Trade	1927	Various	
	278	200	,,	1927		
	391	411	Wolverton	1928/9		
	534	500	Newton Heath	1930		
	Sample Numbers 231331, 235329, 263301, 286040, 300742, 315583, 317842.					
1949	969	150	Derby	1936	725000–725149	Tare
	980	200	Met Cammell	1936	725150–725349	8t 15c
	981	200	Chas Roberts	1936	725350–725549	Code
	982	100	B'ham C & W	1936/7	725550–725649	DBO
	1139	30	Hurst Nelson	1938	725650–725679	
	1140	60	Met Cammell	1938	725680–725739	
	1141	60	Chas Roberts	1938	725740–725799	
2067	1315	200	D'by & W'ton	1942	725800–725999	Tare 10t 15c Code DOUBLE

Diag	*Lot*	*Qty*	*Built*	*Year*	*Numbers*	*Remarks*
2029	313	101	Derby	1927	Various	Tare 7t 17c 3q
	315	100	Wolverton	1927		Code DB
2105	1384	200	Wolverton	1945	726000–726199	Tare
	1414	150	,,	1946	726200–726349	10t 7c
	1458	50	,,	1947	726350–726399	Code
	1525	100	,,	1948	726400–726499	DOUBLE
	1536	550	,,	1949	726500–727049	
2105	Converted to Trestle Plate 726007, 726044, 726054, 726149, 726213, 726223, 726322, 726338, 726429, 726438, 726448, 726489, 726499, 726531, 726579, 726633, 726646, 726699, 726751, 726910, 726983, 726987.					
1674	Converted to Canopy Wagon 321371.					

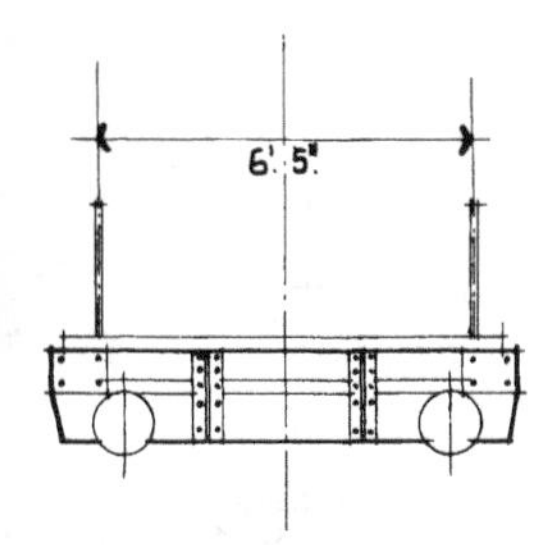

Spoked wheels.
Body, solebars. Grey.
Underframes. Black.
Lettering, Wheel tires. } White.

Fig 22 D1674 Double bolster wagon

The final batch (diagram D2105) were of the all-steel type, carrying capacity being 22 tons. Again, apart from the bolsters, they were identical to the long low wagons to D2083. Like D2067 it was stated that the bolsters were removable and these wagons could again play a dual role. Some vehicles to D2105 were fitted with trestles, thus enabling them to be loaded with plates on edge, which allowed a greater width of plate than could be accommodated when laid flat. Another modification involved D1674; this entailed the fitting of a triangular support at one end, and high stanchions at the other to enable pipes to be loaded, and is illustrated in plate 16D.

Livery Details

The vehicles to D1674 were at first painted grey, and considerable variations occurred. Company initials in all cases extended over the whole height of the side, and were thus about 8 to 9in in height. These were spaced well apart between the bolsters.

Tare weight (with the word tare in full), carrying capacity (in full) and running number were usually placed in line at the left hand end, though some vehicles only had the running number with tare weight beneath it. Non-common-user markings, though not universal, were in the correct position at each end when applied, while some vehicles had the running number painted on the side of the bolsters.

The vehicles to D1949 appear to have come out in a mixture of grey or bauxite livery; the grey vehicles still had the large company initials, but they were placed close together about the centre line. The word DOUBLE, also in large characters, was painted at the right hand end, while the tare weight was also at this end. The running number was at the left hand end, with carrying capacity (shown as 20T) to the right of it.

The bauxite vehicles were laid out in a similar manner; the company initials however were moved towards the left hand end and these and the word DOUBLE were reduced to 4in in height. The later vehicles had the carrying capacity, company initials and running number at the left hand end in line in that order, with the word DOUBLE above and between the first two mentioned items.

Twin Bolster Wagons

THESE vehicles were, in effect, two single bolster wagons, permanently coupled together. They were obtained by modifying mineral wagons of the all-wood type, which were probably of the D1671 type described in chapter six.

Two diagrams were involved; the first type (D2050) retained their bodies with the end doors removed. Longitudinal timbers 5in thick × 10in wide were placed inside the sides; these were about 10ft long, and were probably used to strengthen the underframe. A transverse bolster was then placed on the centre line of the vehicle, the height to the top being 6ft 3¾in above rail level. Carrying capacity was stated to be 24 to 26 tons, according to age and tare of wagons while length of load was not to exceed 33ft 0in; code was TB.

The second diagram D2077, was stated to be for case traffic. Conversion involved the removal of the body, and the removal of the buffers at the inner end of each wagon, which were replaced by solid wood blocks known as dumb buffers. A single transverse bolster was fitted to each wagon, 13ft 6in apart as coupled. Carrying capacity was stated to be 18 tons for the pair, the code being TWINCASE, while the tare weight was 12 tons 8 cwt for the pair. Few other details are known about these vehicles, although it is known that 171 pairs to D2077 were converted.

Flat Case Wagons

THESE were also conversions, the mineral wagons to D1671 again appearing to form the basic vehicle. The diagram number was D2080, and the conversion involved the removal of the body and the fitting of two low transverse bolsters at 9ft 0in centres. The code was FLATCASE, and 146 wagons were so converted in 1942/3. Again little else is known about these vehicles, and no pictures exist to show their livery.

Deal Wagons

UNLIKE the bolsters, which carried timber either as felled or sawn, these vehicles only dealt with the latter category. One diagram, D1681, was involved.

Top left: A deal wagon, D1681 is illustrated in plate 17A, about 250 being built by the LMS.

Centre upper: Soda ash was a rather special traffic and examples of D1951, illustrated in plate 17B and D2022 were not well known vehicles; note the black underframe.

Plate 17C, *centre lower,* is another example of a little-known type, D2131 roadstone wagon, while plate 17D, *bottom,* illustrating 'open wagons' for special traffic, is a 36T match wagon for cranes, another example of a service vehicle. A total of 18 vehicles to four diagrams were built.

Summary of Deal Wagons

Diag	Lot	Qty	Built	Year	Numbers	Remarks
1681	351	100	Newton Heath	1928	Various	Tare
	404	50	" "			7t 11c
	710	50	Wolverton	1933		Code DTK
	1023	50	Derby	1937	490000–490049	

Sample Numbers 326379.

The wagons consisted of a steel underframe of 10in deep section, 27ft 0in over headstocks on a 17ft 6in wheelbase. On this was laid a timber floor 2½in thick, with side and end rails which extended only 1⅝in above the floor; they were thus to all intents and purposes a plain flat wagon. They were unfitted, brakes being of the double type, while carrying capacity was 12 tons. (See plate 17A).

Livery Details

With grey livery these vehicles had 9in high company initials painted on the solebars, placed quite close together just to the right of the left hand wheel. Tare, carrying capacity and running number were painted on the bodyside in that order at the left hand end, while non-common-user markings were placed on the body-side at the extreme ends.

With the change to bauxite some of these positions were reversed; running number, carrying capacity and company initials were now on the body at the left hand end, with the tare weight at the right hand end; non-common-user markings were transferred to the solebars at each end.

Soda Ash Wagons

THESE were large vehicles of the all-steel type. Soda ash is relatively light in relation to bulk, and it is this which accounts for the size of the vehicles.

Two diagrams were issued which had several dimensions and features in common. Steel underframes were constructed from 10in deep sections and were 21ft 6in over headstocks on a 12ft 0in wheelbase. The wagons were unfitted and were provided with handbrakes of the 20 ton standard type; no doors of any type were provided but ladders were fitted for access purposes, while tarpaulin bars were also provided. Bodies were of the rivetted type, those of D1951, rated to carry 20 tons being 5ft 2in high inside, while D2022 wagons rated to carry only 16 tons were no less than 7ft 0in high inside. (Plate 17B).

Livery Details

Wagons to D1951 when built carried the grey livery. Company initials were 18in high and were placed slightly over half way up the bodysides. The running number was at the left hand end, with the carrying capacity above it, tare weight being placed at the lower right hand end. Below the letter M was the legend FOR SODA ASH TRAFFIC in 4in high letters, while immediately below this were the traffic instructions in 3in high letters, an example being: TO WORK BETWEEN WARRINGTON & ST HELENS ONLY, in two lines. With the change to bauxite all lettering was in the correct positions, including non-common-user markings which were absent on the earlier examples; the instruction markings were placed somewhat lower on the side, the first line being amended to FOR SODA ASH.

Roadstone Wagons

THESE vehicles are of interest, inasmuch as they revived a system used many years before by several companies for some coal wagons. These latter had the body formed of several separate containers or boxes, which could be lifted by a crane fitted with a suitable lifting beam and tipped directly where they were required. The vehicles that concern us here had four such containers, of steel construction and a carrying capacity of 3 tons each. The underframes were of the standard steel type on a 10ft 0in wheelbase.

The first diagram (D2113) was for one prototype vehicle, which employed a secondhand underframe previously part of a fish van (No 39373 passenger stock series). This had 3ft 6½in diameter wheels, vacuum pipe and Morton type brakes with two shoes to each wheel. The production batch to D2131 had normal goods stock wheels and Morton brake only, but were otherwise identical to D2113. (Plate 17C).

Summary of Soda Ash Wagons

Diag	Lot	Qty	Built	Year	Numbers	Remarks
1951	971	10	Derby	1936	689000–689009	Tare 9t 2c 3q
2022	1155	50	Derby	1939	689100–689149	Tare 10t 8c

Summary of Roadstone Wagons

Diag	Lot	Qty	Built	Year	Numbers	Remarks
2113	1435	1	Derby	1946	688000	Tare 8t 14c
2131	1489	100	Derby	1947/8	688001–688100	Tare 7t 9c

Wt of skip 9c 3q 71b Cub cap 77 cu ft.

Livery Details

These vehicles had bauxite livery when built; all insignia were on the solebar with the exception of the 'return to' instructions which were on the containers themselves.

Match Wagons For Cranes

THESE vehicles were used to support the jibs of breakdown cranes when travelling, and also to carry tools and equipment.

The first two diagrams were for use with 36 ton cranes; they had steel underframes of 10in deep section, and 15ft 6in wheelbase. D1836 was for wagons 28ft 0in over headstocks, while D1998 was for those 26ft 9in long. The lower edge of the underframe was only 2ft 1in above rail level, and to maintain the buffers at the normal height they were mounted on angle iron frames secured to the top of the underframe; because of the low height 2ft 8½in diameter wheels were fitted.

A box type of jib rest was secured across the frames roughly on the longitudinal centre line, and between this and one end an open body formed from three planks was attached to provide space for tools etc; a note on the diagram stated that such tools were not to exceed two tons in weight. These vehicles were fitted with a hand brake of the double lever operated type, and in addition were vacuum piped, see plate 17D.

The other diagrams covered vehicles for use with 30 ton cranes, and were basically identical. Steel underframes of 10in deep sections were used 33ft 0in over headstocks, and mounted on six wheels at 9ft 6in + 9ft 6in centres. An open-angle iron jib-supporting cradle was provided, its position being the only major difference between the two diagrams. On D2081 this was given as 18ft 8¼in from buffer head to centre line of cradle for use with Cowans Sheldon cranes. On D2082 this dimension was quoted as 14ft 2in, for Ransome & Rapier cranes. These vehicles were not fitted, but had screw couplings and a screw-down hand brake.

Livery Details

Examination of available photographic evidence suggests the use of more than one livery style and it is the Authors' opinion that some were completed in an overall black finish while others had black underframes and below, with probably a dark slate grey body.

Summary of Match Wagons for Cranes

Diag	Lot	Qty	Built	Year	Numbers	Remarks
1836	600	6	Derby	1931	299850–299855	Tare 9t 3c
1998	1069	1	Derby	1937	770000	Tare 8t 15c
2081	Pt1324	5	Wolverton	1942	770006–770010	Tare
	1336	1	,,	1942	770011	10t 13c
2082	Pt1324	5	Wolverton	1942	770001–770005	Tare 10t 13c

CHAPTER 8

VANS FOR SPECIALISED TRAFFIC

THE vehicles described in this chapter were kept strictly to the traffic for which they were designed. They include a high proportion which were fully fitted, and since in many cases their contents were perishable, in general they were used principally on trunk route express goods trains which often ran at night.

Banana Vans

BANANAS were dealt with at few ports, one of the most important being Avonmouth at the south-west extremity of the LMS in the Bristol area. The vehicles for this traffic were confined to very few routes, and special trains were often made up exclusively of such vehicles. Until recent years the

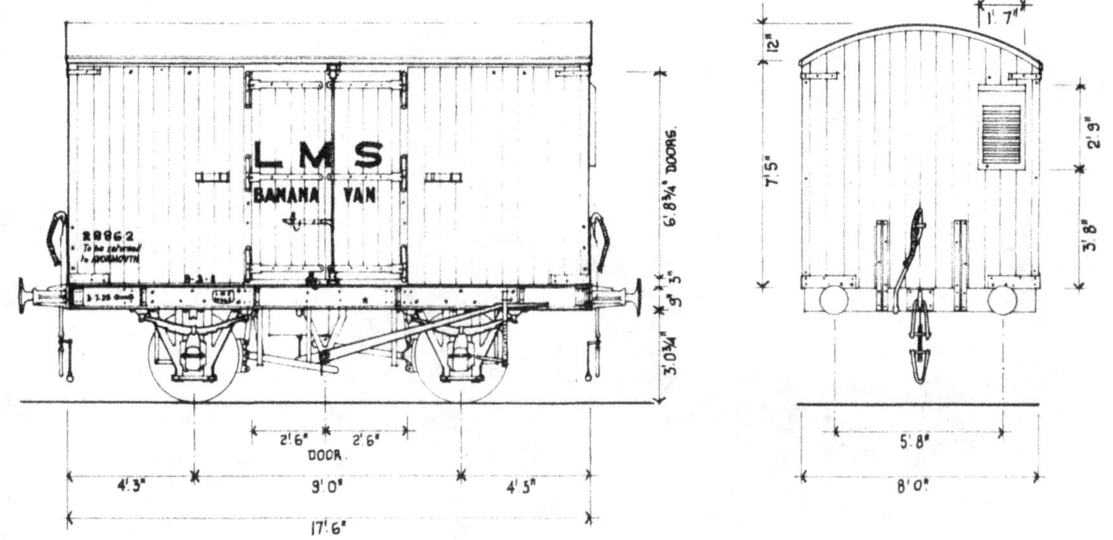

Fig 23 D1660 Banana van

practice was to unload the bananas at the docks in an unripe condition, the final ripening process taking place during the rail journey. The banana vans employed were of special construction, with all-wood double skinned bodies, the outer skin being formed of vertical planks while the inner skin was horizontally planked. The doors were of the side hinged type, double skinned and flush with the sides; to assist the ripening process steam heating coils were provided in the roof space, fed from steam from the engine through train heating pipe connections with which the vans were fitted.

The LMS issued two diagrams for these vehicles; both had standard steel underframes and rather surprisingly, in view of the higher speeds at which these vehicles were worked, both had only a 9ft 0in wheelbase, carrying capacity being 10 tons.

The first diagram (D1660) showed a single louvred ventilator at each end; the vehicles were fully fitted and had Morton type brakes with a single shoe for each wheel. Vehicles to the second diagram (D2111) were unventilated and had a prominent casing over the steam heating pipe at each end; they were also fully fitted and had Morton type brake gear with two shoes for each wheel. (Plate 18A & fig 23).

Livery Details

The vehicles to D1660 at first had 12in high company initials painted on the doors, the L and most of the M being on the left hand door; they were placed just above the middle door hinges, while just below these hinges were the words BANANA VAN, the word BANANA being entirely on the left hand door. The running number was carried at the left hand end of the body about 12in above the bottom edge, while below it was the routing instruction, for example '*To be returned to* AVONMOUTH', the word AVONMOUTH being in block capitals, while the remainder was in italics. The tare weight was to the right of the running number at the bottom edge of the body, while non-common-user markings were placed at each end of the body about 6in from the bottom edge. In addition some vehicles carried a 12in X, denoting their fitted status, on the right hand door about 2ft from the lower edge.

With the bauxite livery the standard layout of insignia was adopted, although the non-common-

Summary of Banana Vans

Diag	*Lot*	*Qty*	*Built*	*Year*	*Numbers*	*Remarks*
1660	204	100	Newton Heath	1925/6		Tare 8t 17c
	206	150	Derby	1925		Code BNV
	222	150	,,	1926		
	239	300	Newton Heath	1926/7		
	307	100	Derby	1927		
	452	100	,,	1929		
	495	66	,,	1930		
	496	34	,,	1930		
Sample Numbers 7888, 14284, 212740, 265063, 268504.						
2111	1421	100	Wolverton	1946	570000–570099	Tare 9t 2c.

Plates 18 and 19 deal with vans or covered vehicles for special traffic. *Top:* Plate 18A illustrates an example of D1660 being repainted before entering traffic in 1945. The dates for painting and lifting can be seen on the solebar, 4.10.45 and the photograph is noteworthy for it clearly shows one method for lettering goods rolling stock. Note the grey patch on the sole-bar giving district number and date of oiling.

Centre: Plate 18B shows the original livery for Refrigerator Vans and it is worth drawing attention to the use of split spoke wheels on the Banana Van and three-hole disc wheels on the Refrigerated Van.

Bottom: Plate 18C illustrates one of these vans in the early BR livery, the LMS livery styles being described in the chapter.

user markings do not appear to have been used, and the words STEAM BANANA were placed above the other insignia at the left hand end, the inscription being in two lines. It should be noted that bananas were not imported into England during the second world war, and these vehicles were used for other traffic during this period. A process of refurbishing them after the war probably ensured that most received the bauxite livery.

Refrigerator Vans

THESE vehicles were built mainly for the transport of imported frozen meat. Like the banana vans they spent much of their life working between the ports and the major distributing centres. Their duties were largely taken over by containers in later years, and, as a direct consequence, none were built after 1930. The history of the LMS vehicles is somewhat complicated, even though only one main diagram was involved. These vehicles to D1672 (plate 18B) had standard steel underframes with a 9ft 0in wheelbase. They were either piped or fully fitted, in both cases being provided with Morton type handbrakes, with one brake shoe to each wheel. Carrying capacity was quoted as 8 tons when running in goods trains, or 6 tons if used in passenger traffic.

The basic body shell was identical to that described for the banana vans, but was unventilated, the space between the outer and inner skins being filled with insulating material, while the inside was lined with zinc to a height of 3ft 0in above floor level. Rails were fitted at intervals across the vehicle at cantrail level, from which carcases could be hung on hooks. To provide refrigeration, ice tanks were fitted across the vehicle inside each end, filled through mounted roof hatches two to each tank. Access to the hatches was obtained by a ladder mounted off centre at one end.

Before a journey the tanks of these vehicles were filled with ice, and this naturally melted as the journey proceeded, causing a certain amount of mess which had to be cleaned up before the vehicle could be used again. With the advent of the so-called 'dry ice' (solid carbon dioxide), which changes directly from the solid to a gaseous state, it was no longer necessary to fill the tanks; instead the 'ice' was placed in with the meat in bags hung from the meat rails. The tanks thus became redundant and were removed from the vehicles as they passed through the shops, the ladders also being removed, though cases are known where the latter were retained at first, but no longer fulfilling their original need. The vehicles modified in this manner were re-classed as Insulated Vans, the diagram number being amended to D1672A.

A second diagram (D1673) described the vehicles to which it applied as Ventilated Refrigerator Vans. They were practically identical outwardly to D1672 but had single louvre ventilators at each end, as on the banana vans to D1660. This diagram was subsequently renumbered D1883, and the vehicles were transferred to the non-passenger coaching stock diagram book. As such they are outside the scope of this book, and are mentioned only to show how close the dividing line was between the two categories.

Finally in this section comes a scheme whereby vehicles to D1672A were to be modified to banana vans. This was dated 1936, and the drawing shows that the resultant vehicles would have been identical to D1660. It was stated that 100 vans were to be so modified, but since they did not carry block numbers, it is not possible to identify the vehicles or to state positively that the scheme was carried out.

Livery Details

These vehicles at first had 18in high company initials, the placing of which was somewhat unusual. The M was placed on the door, and dependent on the type of door fastening employed was either on the left or right hand door. The L

Summary of Refrigerator Vans

Lot	*Qty*	*Built*	*Year*	*Remarks*
5	100	Derby	1924	Tare (piped, 10t 3c
38	100	,,	1925/6	(fitted) 10t 8c
137	50	,,	1926	Note that these tare weights
203	100	,,	1926	are shown on both D1672
295	100	,,	1926/7	and 1672A but are obviously
414	100	Newton Heath	1929	wrong for the latter
451	100	Derby	1929	
499	66	,,	1930	
500	34	,,	1930	

Lots 203 onwards only are shown in official records as being modified to D1672A.
Sample Numbers 7703, 189544, 264756.

and S were positioned equidistant from the M and were thus offset in relation to the ends of the van. The initials were placed about 18in above the lower edge of the body, and about 14in above them was painted the word REFRIGERATOR, or later on (D1672A) INSULATED, these words being about 6in high. In the former case this word was also offset, being slightly closer to the right hand end.

The running number was at the left hand end, with tare weight along the lower edge of the body just to the left of the door; fitted examples also carried a 12in X on the lower edge of the door. Although we have not located any photographs of these vans in the later livery, it is likely that they followed the standard layout, with the word INSULATED above the other insignia at the left hand end.

Ventilated Meat Vans

THESE vehicles were used for the carriage of fresh meat, which unlike the frozen variety merely had to be kept cool during transit. The construction of these vans, therefore, was very similar to that of the ordinary covered goods, including sliding doors, and incorporating additional ventilation to ensure a cool interior.

All were built on standard steel underframes, and were either piped or fully fitted, handbrakes being of the Morton type with a single shoe on each wheel. Carrying capacity was 8 tons in goods or 6 tons in passenger traffic.

The first diagram (D1670) had a 9ft 0in wheelbase. The body was of the all-wood type, with horizontally planked sides and vertically planked ends and doors. Adequate ventilation was ensured by having single hood type ventilators on each end, four torpedo type ventilators on the roof, and two large louvre type ventilators in each side.

The other two diagrams in this group specified corrugated steel ends of the early type, with vertically planked sides and doors. The louvre type ventilators were dispensed with, and to compensate for these, twin hood type ventilators were fitted at each end, while the roof ventilators were also retained. D1821 had a 9ft 0in wheelbase, while on D1822 this was increased to 10ft 0in, and 3ft 6½in diameter wheels were fitted. Finally it should be noted that in addition to these details, these vehicles were provided with rails from which the meat could be hung in a similar manner to that already described for the refrigerator vans (plate 18C).

Livery Details

Very few photographs exist of these vehicles in LMS livery and those available are in the early livery style with a grey body.

Two styles are apparent and both have the word MEAT at the top of the door in letters approximately 12in high. The first style used the letters LMS spaced across the body with the L covering the 4th, 5th and part of the 6th plank located equidistant between the vertical strapping. The M was in the centre of the door and the S at the extreme end of the vehicle. The running number was at the left hand end on the second plank and beneath this was written MEAT TRAFFIC/ RETURN TO INVERNESS with N to the left of the inscription. A large X was beneath the M and the tare weight was on the solebar at the right hand end of the vehicle.

Later the LMS was placed on the door beneath the word MEAT with the running number still on the second plank and the non-common-user markings on the bottom plank; however, with this style N was used at both ends. It is felt that in bauxite a layout similar to the style depicted in plate 18C was used, incorporating the non-common-user markings as before.

Beer Vans

THERE is little need to describe the contents of these vehicles, except to say that the liquid in question was conveyed mainly in barrels. Again the wagons kept to a few selected routes, one of the best known being from Burton-on-Trent to St Pancras. The foundations of the latter station were arranged as vaults, with the supports spaced to accommodate the maximum number of such barrels.

Only one diagram was involved, No D1817. Ventilation was again of prime importance, and the all-wood body was built of horizontal planks with a pronounced gap between them. Solid sliding doors of the vertically planked type were

Summary of Ventilated Meat Vans

Diag	*Lot*	*Qty*	*Built*	*Year*	*Remarks*
1670	306	300	Wolverton	1927	Tare (piped) 8t 6c (fitted) 8t 13c
	480	100	″	1930	Code (piped) PMV (fitted) FMV
Sample Numbers 173127, 267107.					
1821	497	66	Derby	1930/1	Tare (fitted) 8t 10c
1822	498	34	Derby	1930/1	Tare (fitted) 9t 2c

Top: Plate 19A calls for little comment, it is one of the few known photographs of a beer van. They usually ran either in block trains or comprised a good section of the train.

Centre: Plate 19B illustrates a gunpowder van with the post-1936 livery; the 'Instanter' couplings, usually associated with the GWR, can clearly be seen.

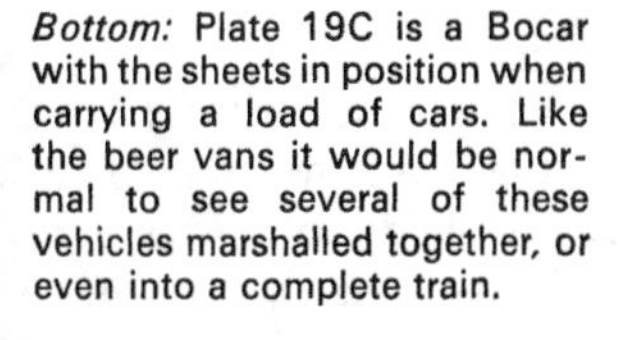

Bottom: Plate 19C is a Bocar with the sheets in position when carrying a load of cars. Like the beer vans it would be normal to see several of these vehicles marshalled together, or even into a complete train.

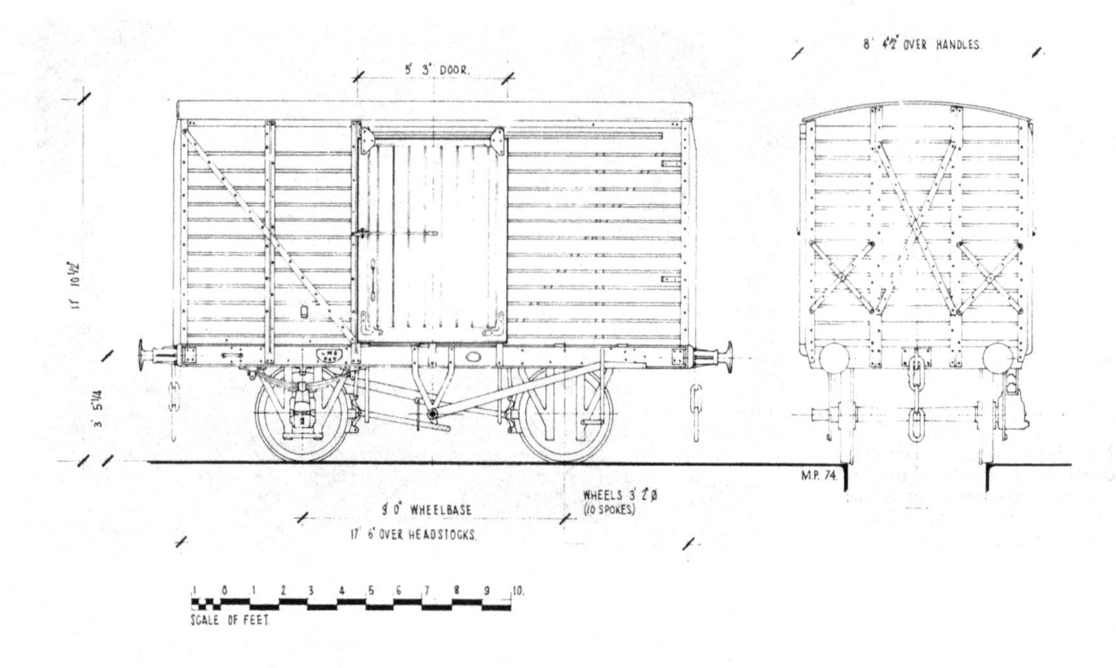

Fig 24 D1817 Beer van

Summary of Beer Vans

Diag	*Lot*	*Qty*	*Built*	*Year*	*Numbers*	*Remarks*
1817	419	100	Derby	1929	189331–189430	Tare 7t 4c Code BRW

provided in each side. The vehicles were unfitted, standard steel underframes with a 9ft 0in wheelbase and Morton type handbrakes being provided, while carrying capacity was quoted as 12 tons. (Plate 19A and fig 24).

Livery Details

The initial livery was as shown in plate 19A; company initials were 12in high, the words BEER VAN being about 5in high. Although we have been unable to trace a photograph of one of these vehicles in bauxite livery it is probable that the standard layout was followed as closely as possible, with the words BEER VAN above the other insignia at the left hand end.

Gunpowder Vans

In view of the dangerous nature of their contents, it is not surprising to find that these vehicles were of all-steel construction. Two diagrams were issued by the LMS for these vehicles which externally appear identical, the only difference being that D1665 fig 25 were rated to carry 7 tons, while D2093 was rated at 11 tons.

These vehicles were somewhat smaller than other contemporary vans, length over headstocks being 16ft 6in on a 9ft 0in wheelbase, while overall height was only 10ft 7in. The body was built of steel plates rivetted together, with a number of tee section uprights to give rigidity; inside was a wood lining. The roof was also of steel plates, and no ventilation was provided. These vehicles were unfitted, and were provided with a double hand brake, *Instanter* couplings being provided in many cases. (Plate 19B).

Livery Details

Plate 19B clearly shows the layout of livery for a gunpowder van built in 1937; this was one of the last vehicles built to use grey body colour with the later livery style normally associated with bauxite body colour. Apart from the body colour change, the style was used for lots 923 onwards and it is interesting to record that although the change to bauxite was supposed to be made in May 1936, vehicles recorded as being built in 1937 still retained the early body livery. No photographs of these vehicles in early LMS livery are known to exist but the probable livery would be GUNPOWDER, GUN to the left of the door, POWDER on the door, L beneath the GUN and MS on the door. GUNPOWDER was probably 12in

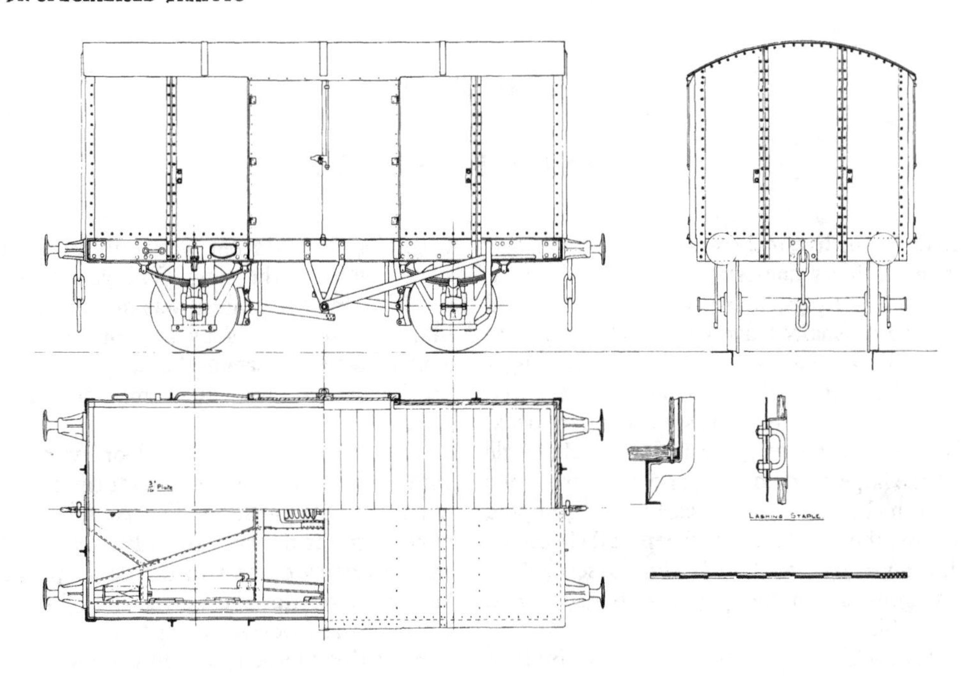

Fig 25 D1665 Gunpowder van

Summary of Gun powder Vans

Diag	Lot	Qty	Built	Year	Numbers	Remarks
1665	27	10	Earlstown	1923/4		Tare 8t 1c
	109	40	,,	1925		Code GPV
	415	25	,,	1929		
	709	20	Derby	1933	Sample 299031	
	923	20	,,	1936	701000–701019	
	1027	25	,,	1937	701020–701044	
	1200	20	,,	1939	701045–701064	
2093	1337	20	Wolverton	1943	701065–701084	Tare 8t
	1349	20	,,	1944	260928–260947	Built for LNER
	1474	15	,,	1948	701085–701099	

high and LMS 18in high. The running number was on the extreme left hand end and the tare weight at the extreme right end.

Bocar

THESE vehicles were built to transport car bodies from the press shops to the factories where they were completed. They looked, and were, of somewhat flimsy construction. The underframe was of typical coach construction 53ft 6in over headstocks (number 700931 was 50ft 0in long), carried on two 8ft 0in wheelbase bogies set at 39ft 0in centres.

The body as can be seen from plate 19C, consisted of horizontally planked ends with a light open framing between them to give stability. Tarpaulin sheets were arranged to provide cover for the contents and were secured to the underframe by short lengths of rope. These vehicles were fitted, and had a handbrake which was operated by a very short lever.

Livery Details

The livery style is depicted in plate 19C, the only point to be noted being that the running number and code were repeated along the longitudinal rail.

Summary of Bo Cars

Diag	Lot	Qty	Built	Year	Numbers	Remarks
2114	1487	1	Derby	1946	700900	Tare 19 tons to
	1496	35	,,	1947/8	700901–700935	20t 8c Code Bocar 700931 tare 18t 11c

CHAPTER 9

LIVESTOCK VEHICLES

THE majority of livestock carried by the railways was of the ordinary domestic kind, namely horses, cattle, sheep and pigs. Exotic animals of the kind found in zoos usually travelled, suitably caged in passenger brake vans; horses were also passenger rated and horseboxes therefore were to be found in the non-passenger coaching stock diagram book.

Of the other three categories of livestock, cattle formed the largest proportion, and apart from one or two vehicles built in the early post-grouping period for the carriage of sheep, all livestock vehicles built by the LMS were described as cattle wagons, sheep and pigs also being carried in these vehicles as required.

The first vehicles to be described were built to an LNWR drawing dated 1910; they were of the all-wood type and were 18ft 9in long over headstocks on a 10ft 0in wheelbase. The body had a heavy external wooden frame, the sides being horizontally planked for two thirds of their height, the bottom plank having a gap above and below. The ends were planked over their whole height, with two heavy wooden stanchions to provide stiffness; doors provided in each side and were in three parts. The top part consisted of two side hinged doors, while below was a bottom hinged door which provided a ramp between the vehicle and loading dock. The doors opened to the full height of the sides but were only planked to the same height as the sides. The gap between this planking and the cantrail was divided horizontally into two equal parts by an iron bar; while the animals were thus assured of ample air they could not be injured through poking their heads over the sides.

The floor and drop flap doors were provided with battens to give a foothold for the beasts; inside was a movable partition. This could be positioned either right at the end, roughly mid-way between end and door, or near to the door opening. This helped to restrict the movement of the animals when a part load was being carried. Handbrakes were of the double type, with the long side lever favoured by the LNWR, while the drawing also shows a vacuum pipe.

The first vehicles to be allocated LMS diagram numbers were actually of MR design, the first lot being included among that company's stock. The LMS issued two outwardly identical diagrams for these vehicles, the first (D1661, fig 26) being built between 1922–31, while the second (D1840) was built between 1932–4. The main reason for the two diagrams appears to be that the second was shown as fitted, while the first were either unfitted or piped only.

These vehicles were of all-wood construction

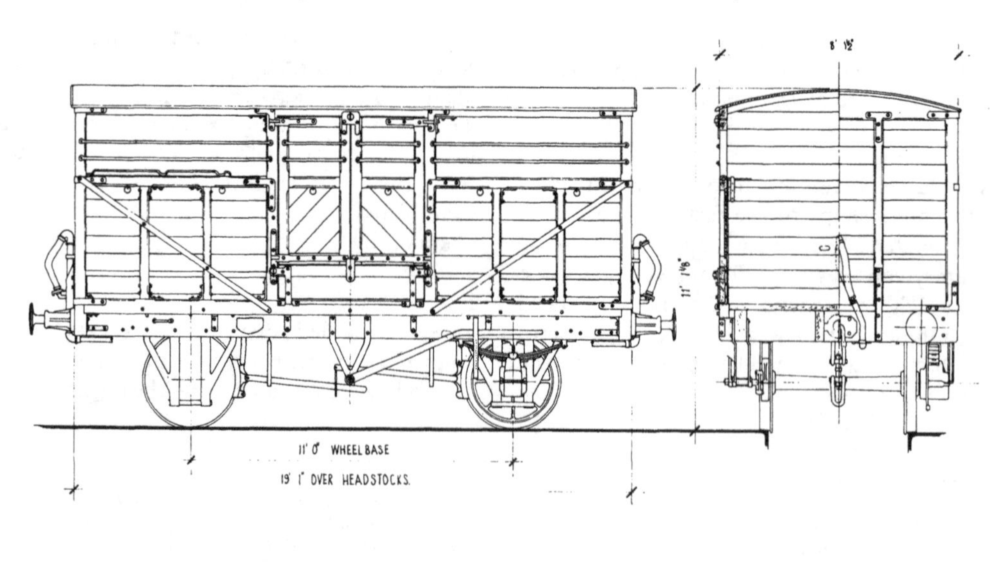

Fig 26 D1661 Cattle wagon

Plates 20A, *above*, and 20B, *below*, show both styles of LMS cattle wagon and different liveries, details of which can be clearly seen.

Cattle wagons in their latter days were used, and indeed branded accordingly, for other traffic, eg ALE WAGON.

19ft 1in over headstocks on an 11ft 0in wheelbase. Construction was basically the same as that already described for the LNWR vehicles, but included vertical intermediate bodyside stiffeners rather than the diagonal type employed on the former. Many small variations were to be found on these vehicles, particularly regarding the drop door, which on some vehicles was much shallower than on others. The handbrakes on D1661 were of the double type, while on D1840 they were of the Morton pattern, and the sample numbers in the summary are based on the latter types. (Plate 20A).

The last LMS diagram for this class of vehicle was D1944; only one lot was built by the LMS, appearing in 1935, but like the LNWR vehicles with which we started this chapter, history repeated itself and identical vehicles appeared under BR auspices. These vehicles had all-wood bodies, on which the outside wooden framing was replaced by angle irons, but which were otherwise very similar to those already described. The underframes were of the steel type, 18ft 6in over headstocks on an 11ft 0in wheelbase. They were fully fitted and had handbrakes of the Morton type, with twin brake shoes on each wheel. All of the LMS standard vehicles had a carrying capacity of 12 tons. (Plate 20B).

As already mentioned the LMS also built one or two vehicles which did not fall into the normal cattle wagon class; of these the first diagram D1818 was for a single convertible sheep and cattle wagon. This vehicle had a steel underframe 19ft 0in long over headstocks on an 11ft 0in wheelbase. Hand brakes of the double type were provided, and the vehicle was also piped.

The bodysides were horizontally planked, wide planks being used for the lower two thirds of the height and narrow planks for the upper third, gaps being left between them in both cases. Doors of the normal cattle wagon type were provided, with guard rails across the upper portion, while the ends were horizontally planked without any gaps. A false floor was provided, which was carried in the roof space when the vehicle was being used for cattle, and which could be lowered to provide two roughly equal compartments one above the other when required for sheep. This operation was effected by means of steel ropes which passed over pulleys to a die block mounted on a screw operated by detachable handles. Another unusual point about this vehicle was that the roof was of corrugated iron. Carrying capacity was 12 tons.

The second diagram D1824 was also for a single vehicle, in this case a double deck sheep truck. Construction was very similar to that of D1818, but the vehicle was mounted on a wooden underframe 19ft 1in over headstocks, wheelbase being 11ft 0in. The only other points where this vehicle differed from the former lay in the roof which was of normal wooden construction; the false floor which in this case rested on the main floor was

Fig 27 D1819 Calf van

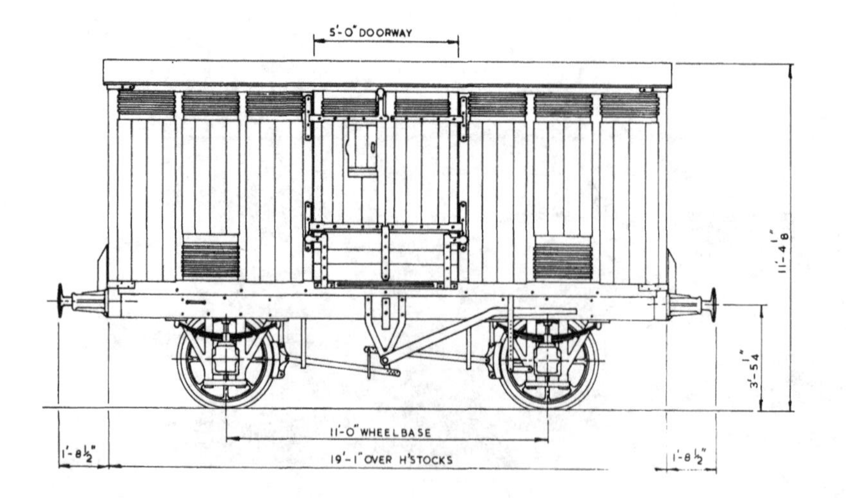

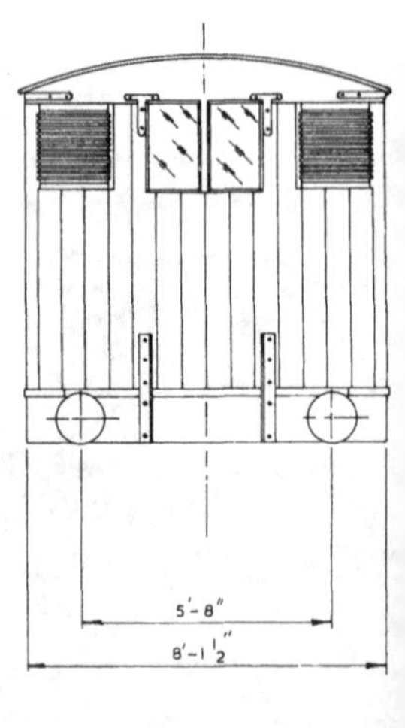

raised into its midway position. Carrying capacity was again 12 tons.

The final diagram in this group was less experimental in character, and was a modified form of an MR design (D1819, fig 27). This was for calf vans, which were very similar to prize cattle wagons which were counted as non-passenger coaching stock. They were of all-wood construction, length over headstocks being 19ft 1in on an 11ft 0in wheelbase. The body was double skinned, vertically planked, and was totally enclosed. The doors were again in three parts, the two upper leaves being of the side hinged type, with a drop flap door below them. Ample ventilation was ensured by the provision of numerous louvre type ventilators, while two glazed windows were provided at the top of each end. Hand brakes of the double type were fitted, and the vehicles were also piped. Carrying capacity was again quoted as 12 tons.

Livery Details

There are no known photographs of calf vans in LMS livery; indeed the only photograph ever seen by the authors was one, No. 327099 in BR ownership, branded FRUIT! It is probable that when first built a livery based upon MR style was used and we feel that 'Calf Van' would have appeared high up on the doors with LMS spread across the body, the M being 'off centre' and to the left. The running number would probably be at the left hand end while, by no means certain, but quite possible, the legend: 'To be returned to Derby Cattle Docks when empty,' could have been at the extreme bottom right hand end of the vehicle.

The LMS did not build any cattle wagons after the change in livery style, so only repaints after 1936 would have appeared in the bauxite body colour with the LMS etc, in the standard position.

Summary Cattle Wagons

LNWR Design to Drawing Number 60

Lot	*Qty*	*Built*	*Year*
4	90	Earlestown	1924
26	400	"	1924
DIAG 1661			
Lot	*Qty*	*Built*	*Year*
987*	300	Derby	1922
3	150	"	1924
106	250	"	1924/5
142	200	"	1925
152	400	"	1925/7
193	100	Trade	1925/6
194	100	"	1925
195	50	"	1925
196	50	"	1925
197	100	"	1925
198	100	"	1925
238	250	Newton Heath	1926

Lot	*Qty*	*Built*	*Year*
279		Trade	1927
280		"	1926/7
281		"	1927
282	500 total	"	1927
283	for lots	"	1927
284	279/86	"	1927
285		"	1927/8
286		"	1928
310	200	Newton Heath	1927
329	200	" "	1927
368	500	" "	1928
437	589	Derby	1929/30
501	250	"	1930
538	100	"	1931
539	49	"	1931

*MR Lot 136 piped 164 unfitted
Code (unfitted) C Tare 7t 14c
(piped) PC 8t 2c
Sample Numbers 12098, 230909, 265032

Diag	*Lot*	*Qty*	*Built*	*Year*	*Numbers*	*Remarks*
1840	652	150	Derby	1932/3		Tare 8t 11c
	653	200	"	1933		Code FC
	711	125	"	1933		
	712	125	"	1933		
	756	50	"	1934		
Sample Numbers 293612, 296640.						
1944	834	100	Derby	1935	710000–710099	Tare Code FC

CONVERTABLE SHEEP AND CATTLE WAGON D1818

Lot 459 Qty 1 Built Derby 1929 Tare 10t 18c

DOUBLE DECK SHEEP TRUCK D1824

Lot 520 Qty 1 Built Derby 1930 Tare 9t 1c

CALF VAN B1819

Lot 473 Qty 10 Built Derby 1929 Code CV Tare 8t 1c

CHAPTER 10

TANK WAGONS

RAILWAY owned tank wagons were a very rare breed indeed, the majority being owned by the petroleum companies. The LMS built no more than about 70 vehicles which fall into this category, but issued no fewer than 18 diagrams to cover them, and they form a most varied and interesting group.

Travelling Gas Holder Trucks

THESE vehicles were used to supply gas to those vehicles which were gas lit coaches, and to kitchen cars which used gas for cooking. The first type, to D1816, was built at Derby in 1926; only two vehicles were involved and they employed secondhand underframes, probably from four-wheeled coaches. Two sets of dimensions were quoted for the underframes:

25ft 0in over headstocks, 15ft 0in wheelbase
20ft 0in over headstocks, 12ft 0in wheelbase

The buffers were 2ft 0in long in both cases, and wheels were 3ft 6in in diameter, the vehicles were fully brake fitted. Twin tanks 18ft 0in long by 4ft 0in diameter were fitted side by side; they were in line with the headstocks at one end, a small platform being left at the other end where the valves were arranged. (Plate 20C).

The second diagram (D1815) was for very similar vehicles; in this case length over headstocks was 21ft 0in on a 12ft 0in wheelbase, otherwise all of the remarks and dimensions above apply equally to these vehicles.

The third diagram, D1831 (fig 28), was for new vehicles. They had steel underframes of 10in deep section, 25ft 0in long over headstocks on a 15ft 0in wheelbase. Wheels were 3ft 7⅞in diameter, and these vehicles were also fully fitted. A single tank 24ft 0in long by 7ft 8in diameter was positioned centrally on the underframe.

The next two diagrams again applied to vehicles built on secondhand underframes, of wooden construction 30ft 5in over headstocks. They had six wheels, wheelbase being 10ft 6in + 10ft 6in, with wheels 3ft 7½in diameter, while the buffers were 1ft 10½in long. The tanks were mounted

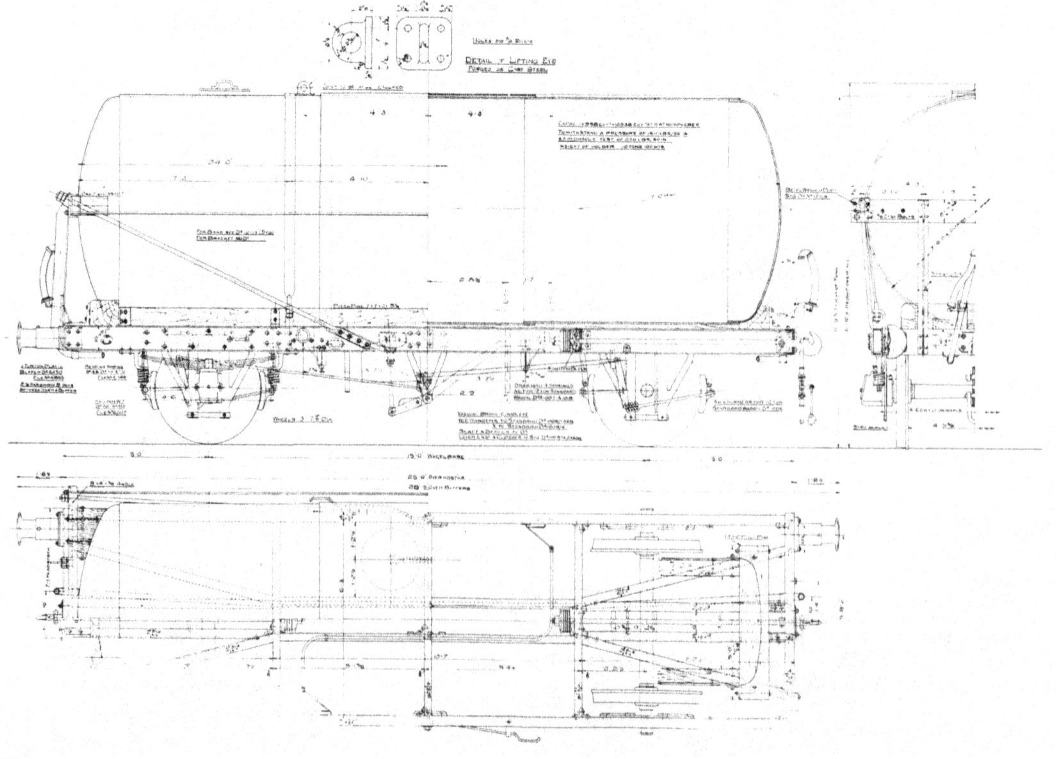

Left: Fig 28 D1831 Gas holder truck

Above: Plate 20C illustrates a typical LMS gas tank wagon, probably of LNWR origin. Note the use of a secondhand underframe, long buffers etc. It has carriage style letters and numbers. Date and location of photograph unknown. R. J. Essery collection

centrally in the longitudinal direction leaving a platform at each end. D1826 had twin tanks, 25ft 0in long by 4ft 5in diameter, mounted side by side, and was fully fitted, while D1827 had a single tank 23ft 0in long by 5ft 6in diameter, this vehicle having hand brakes only.

The two final diagrams in this group were for new construction, the first (D1825) having a steel underframe of 9in deep section, 21ft 0in over headstocks on a 12ft 0in wheelbase. Wheels were 3ft 7½in diameter, and buffers 2ft 0in long; they were unfitted. Three tanks were provided, two side by side at the bottom with the third mounted centrally above them. The tanks were 18ft 4in long by 4ft 0in diameter and were mounted in line with the headstock at one end, leaving a platform at the other end.

The last diagram in this group, D2048, also had a steel underframe, built from 11½in deep sections, length over headstocks being 20ft 11½in on a 12ft 0in wheelbase. The buffers were 2ft 0in long, while the wheels appear to have been 3ft 7½in diameter. These vehicles were fully fitted, the hand brake probably being of the Morton type

Summary of Travelling Gas Holder Trucks

Diag	*Lot*	*Qty*	*Built*	*Year*	*Numbers*	*Remarks*
1816	288	2	Derby	1926		Tare 12t 9c Code GAS
1815	311	3	St Rollox	1927		Tare 13t 6c
	416	2	Derby	1928		
	514	2	"	1930		
	563	3	"	1930		
1831	359	3	Chas Roberts	1928	31897	Tare 17t 9c
1826	567	4	Derby	1930	317279	Tare 17t 7c
1827	568	1	Derby	1930		
1825	566	2	Derby	1930		
2048	1233	4	Derby	1940	748800–748803	Tare 15t 2c

Plate 20D is the only known photograph of an LMS Creosote tank wagon, built by Charles Roberts of Wakefield. The similarity to petrol tank wagons of the period should be noted. Charles Roberts Ltd.

with twin shoes on each wheel. A single tank was fitted 20ft 0in long by 7ft 6in diameter.

Livery Details
Very few photographs exist; the best known to the authors was reproduced in *British Goods Wagons* and clearly shows grey livery, which was also used with D1831 with the legend LMS centrally upon the tank, GAS STORE HOLDER to the right and running number to the left. However, the print from which plate 20C was made suggests coach style livery but it is impossible to be certain; equally it is not possible to state definitely what was the body colour.

Creosote Tank Wagons

THESE vehicles were used for the transport of creosote to sleeper depots. Early vehicles were of the rectangular tank type, but all LMS built examples were however of the conventional round tank type.

All had steel underframes built from 10in deep sections, wagons to diagrams D1820 and 2033 being 16ft 6in long over headstocks on a 9ft 0in wheelbase. The tanks were 15ft 5in long, that on D1820 being 5ft 1in diameter, while D2033 was 5ft 6in in diameter. These vehicles were unfitted, D1820 having double brakes and D2033 Morton type brakes. (Plate 20D).

The final diagram (D2031) was for a vehicle 17ft 6in over headstocks on a 10ft 0in wheelbase, tank dimensions being 17ft 5in long by 6ft $3\frac{3}{8}$in diameter; this vehicle was also unfitted and was provided with Morton type handbrakes.

Livery Details
The only known photograph is reproduced as plate 20D and it is not known what body colour or lettering style was adopted after 1936.

Ammoniacal Liquor Tank

LITTLE is known about this vehicle beyond its diagram number, D2060, and its running number 198608. It was of conventional appearance, with a steel underframe of 9in deep sections which was 21ft 0in over headstocks with a 12ft 6in wheelbase; it was unfitted and had a handbrake of the Morton type. The tank was 17ft 3in long by 5ft $10\frac{1}{2}$in diameter, mounted centrally on the underframe.

Tank Wagons for Cable Compound

EXACTLY what was cable compound is not clear to the authors. What is clear, however, is that the two vehicles to D2030 were very similar to the contemporary milk tanks. The steel underframes of 10in deep section were 20ft 6in over headstocks, and had six wheels of 3ft $6\frac{1}{2}$in diameter at 6ft 6in + 6ft 6in centres; both were fully fitted. The

Summary of Creosote Tank Wagons

Diag	Lot	Qty	Built	Year	Numbers	Remarks
1820	475	9	Chas Roberts	1929	304592	Tare 8t 9c
	1160	1	" "	1938	748900	Code GAS Capacity 2066 gal
2033	1235	2	Chas Roberts	1939	748901–2	Tare 8t 11c Capacity 2066 gal
2031	1236	1	Chas Roberts	1939	748903	Tare Capacity 3062 gal

tank was 16ft 8in long inside by 6ft 4½in diameter, a thick clothing bringing the outside length to 17ft 5in by 7ft 1⅝in diameter. The tank was mounted centrally on the underframe, with an access ladder on one side.

Livery Details
No photographs are known to exist and therefore livery details cannot be given.

Summary of Tank Wagon for Cable Compound

Lot	Qty	Built	Year	Numbers	Remarks
1239	1	Derby	1939	707300	
1492	1	,,	1947	707301	

Tank Wagons for Beer

THESE were conventional tank wagons, mounted on standard steel underframes of 10in deep section, and having a wheelbase of 10ft 0in. They were fully fitted. The tank was 15ft 4in long by 6ft 4⅜in diameter. The diagram number was D2037.

Livery Details
No photographs of these vehicles are known to exist, and therefore livery details cannot be given; it is thought that they may have carried the liveries of the brewing companies concerned.

Summary of Beer Tank

Lot	Qty	Built	Year	Numbers	Remarks
1241	2	Derby	1939	707105–6	Tare 11t 2c
1380	1	,,	1944	707107	
1437	1	,,	1947	707118	

The remaining vehicles in this chapter are not true tank wagons, as the diagrams refer only to the underframes, on which were mounted tanks of various types. Before describing them, brief mention must be made of Lot 202, which was described in the lot book merely as a 14-ton cylindrical tank wagon, built by the trade probably about 1925. Nothing else is known of this vehicle, but it was probably very similar to the creosote wagons previously described.

Demountable Tanks for Loch Katrine Water

LOCH Katrine water is an essential ingredient in the distilling of Scottish whisky. These vehicles consisted of a standard steel underframe, having a 10ft 0in wheelbase. They were fully fitted and had Morton type handbrakes, with twin shoes to each wheel. D2034 carried a single small tank, while D2141 had provision for twin tanks.

Livery Details
The underframes appear to have been painted black, D2034 had the company initials just to the left of centre with the running number to the right of this, the only other marking being SHUNT WITH CARE above the right hand wheel. D2141 had the running number prefixed M near the left hand wheels, and the same SHUNT WITH CARE towards the right hand end.

The tank on D2034 was noteworthy in that it was finished in the GWR coach livery of chocolate and cream at the request of the late Mr Norris who was an enthusiast for that line, and a director of the whisky company. It is not certain how the tanks on D2141 were finished, they were however in two colours and were labelled for Lemon Hart & Son.

Chassis for Road/Rail Beer Tanks

THESE vehicles to D2035 also consisted of a standard steel underframe, having a 10ft 0in wheelbase. They were fully fitted with handbrakes of the double shoe Morton type. The top of the underframe had a steel deck, with hinged flaps which could be extended over the buffers. Longitudinal rails were provided to guide the road wheels of the tanks, which were shown as having four wheels at 7ft 6in centres while chains with turnbuckles were provided to secure the tanks.

Livery Details
We believe that these vehicles were mostly black but the raised edge of the floor was painted bauxite. However, since this was covered with 10T LMS 707109, SHUNT WITH CARE FOR USE OF WHITBREAD'S TANKS ONLY, or similar the vehicles are rather hard to describe. The central ramp also carried detailed loading instructions written in white together with white stars located centrally.

Summary of Demountable Tank for Loch Katrine Water

Diag	Lot	Qty	Built	Year	Numbers	Remarks
2034	1237	1	Derby	1939	707200	Tare 6t 18c
2141	1490	1	Derby	1949	707201	Tare

Summary for Road Rail Beer Tanks

Diag	Lot	Qty	Built	Year	Numbers	Remarks
2035	1230	5	Derby	1939	707100–707104	Tare 9t 3c
	1419	10	Wolverton	1946	707108–707117	Capacity 10 ton

Chassis for Sodium Silicate Tank

THIS vehicle also consisted of a standard steel underframe, having a 9ft 0in wheelbase, it was fully fitted and probably had Morton type handbrakes. It had a timber deck which extended from the headstocks at each end, leaving a gap of about 8ft 0in in the centre. A small tank about 10ft 0in long was fitted centrally on the underframe. The diagram number was D2135.

Summary of Chassis for Sodium Silicate Tank
Lot 1539 Qty 1 Built Derby 1947 Tare 6t 14c

Chassis for Road-Rail Edible Oil Tanks

THESE vehicles to D1988 had a steel underframe of 10in deep section, 30ft 0in over headstocks and mounted on six wheels 3ft 6½in diameter, at 9ft 6in + 9ft 6in centres. They were fully fitted, with two brake shoes to each wheel and a handbrake with a very short lever. They were also fitted with accumulator boxes and dynamos.

Sides built of steel plate, about 18in high were fitted together with hinged flaps which extended over the buffers for loading purposes. The road tanks were of the semi-trailer type, and doubtless were pulled by the mechanical horse type of road tractor. Electric immersion heaters were fitted in the tanks to keep the oil warm, the accumulators and dynamo supplying electricity for this purpose while on the railway. Two tanks could be carried on each wagon, and while it is not known how many of these existed it would seem that there were more than could be accommodated on the wagons at any one time, presumably as some would have been on the road part of the journey.

Livery Details
The livery was generally similar to that of the beer tanks except that the running number, LMS, SHUNT WITH CARE etc, were located upon the solebars.

Summary of Chassis For Road-Rail Edible Oil Tanks
Lot 1079 Qty 8 Built Derby 1938 Numbers 707000–707007

CHAPTER 11

HOPPER WAGONS

THE vehicles described in this chapter cover a large number of designs, built for a variety of traffics. It will be noted that some of the wagons fall into categories already described in chapters six and seven but the vehicles described here had one important feature in common, their ability to discharge their loads entirely by gravity. The majority of hopper wagons were of the all-steel type of construction, with the plates rivetted together, and unless stated otherwise, it can be taken that this form of construction applied to the vehicles described in this chapter.

Ballast Hopper

Two diagrams were issued for this class of wagon, which belonged to the Civil Engineers Department. The first vehicles, to diagram D1800 were built in 1928, and had underframes 21ft 2in long over headstocks on a 12ft 6in wheelbase. They had a carrying capacity of 25 tons, with a capacity of 550 cu ft and tare weight of 10 tons, overall height being 8ft 9⅜in. They were unfitted and were provided with a screw down hand brake. The body was 18ft 2½in long, being offset on the underframe so that it was in line with the headstock at one end. At the opposite end was a platform with three handwheels, arranged across the vehicle, each operating a bottom door 8ft 0in long, placed longitudinally between the axles, and arranged so that the load could be discharged to either side or between the track as required.

The second diagram D1804 was for wagons built in 1932, having steel underframes of 12in deep sections. They were 20ft 6in long over headstocks on a 12ft 0in wheelbase. They had a carrying capacity of 25 tons, with a capacity of 613 cu ft and tare weight of 8 tons 13 cwt; overall height was 7ft 6in. These vehicles were also unfitted and had hand brakes of the double type. (Plate 21A). The body was the same length as the under-

Summary of Ballast Hoppers

Diag	*Lot*	*Qty*	*Built*	*Year*	*Numbers*
1800	371	20	Leeds Forge	1928	282504
1804	634	108	Met Cammell	1932	197272–197379

Above: Plate 21A clearly shows the livery of a steel hopper ballast wagon and is undoubtedly grey in colour. See chapter two for the colour of service stock. It was photographed on 10 March 1932.

Below: Plate 21B is an ex-works photograph of a trade-built cement hopper.

frame, and was provided with four longitudinal bottom doors, 4ft 10in long, operated by hand-wheels on the solebars.

Livery Details
Not many photographs exist showing these vehicles but on D1800, (with the higher sides than D1804), the livery adopted was to place the LMS, with each letter some 18in high, between the angle irons, with ED located beneath the M and S on the underside of the hopper, and the running number as on plate 21A. It is presumed that after 1936 the small size livery was adopted, located probably at the left hand side.

Cement Hopper

ONLY one diagram was issued for these vehicles, D1806. They were built in 1932, and were 19ft 9in long over headstocks on a 12ft 0in wheelbase. They were coded HPO, the O indicating the carrying capacity, 20 tons, but no cubic capacity was quoted. Tare weight was 10 tons 18 cwt, and overall height 11ft 1⅛in. They were unfitted, and although the diagram states that they were provided with double brakes, photographs show that some at least had what appears to be the Morton type, but with brake shoes on all wheels. As cement had to be kept dry, these vehicles were covered, and provided with three loading doors in the roof, while four bottom discharge doors were fitted, operated by simple hand levers. (Plate 21B).

Livery Details
The only known photograph has been used as plate 21B.

Summary of Cement Hopper

Diag	Lot	Qty	Built	Year	Numbers
1806	639		Met Cammell	1932	299894–

Loco Coal Hopper

Two diagrams were issued for these vehicles, both of which were of the all-wood type of construction. The bodies were of the seven plank type, and externally there was little to distinguish them from normal mineral wagons, apart from the lack of side or end doors. Internally they had sloping floors forming the familiar hopper shape.

Wagons to the first diagram (D1668) were built in 1925, and were of LNWR design. Length over headstocks was 18ft 0in, on a 9ft 9in wheelbase. They were coded HPM the M indicating a carrying capacity of 15 tons, while capacity was 585 cu ft and tare weight 7 tons 19 cwt, overall height being 8ft 3½in. They were unfitted, the diagram stating that hand brakes of the double type were provided. Six bottom doors were provided four, between the axles being hinged along their outer edges, while the remaining two, above the axles, were hinged across the wagon at their inner ends, thus protecting the axles when opened.

Wagons to the second diagram (D2024) were built in 1939, and were not specifically branded as 'loco coal'; they are included here as it seems likely that they were used for this purpose. They conformed to the normal mineral wagon dimensions, namely 16ft 6in over headstocks on a 9ft 0in wheelbase. They were coded HPW; in this case the W probably indicated 'wood' as this letter was not used to indicate carrying capacity which in this case was 10 tons, capacity being 400 cu ft and tare weight 8 tons 1 cwt. Overall height was 8ft 6¼in, and handbrakes of the double type were provided. Four bottom doors were provided, hinged along the longitudinal centre line of the wagons, all being between the axles. (Plate 21C).

Livery Details
Again the only known photograph has been used in plate 21C and it is presumed that those built before 1936 were Delivered with a large centrally placed LMS and LOCO as in plate 16A.

Bogie Coal Hopper

THESE vehicles to D1708 were built in 1929, specifically to carry coal from the pits to the LMS power station at Stonebridge Park in north-west London, which supplied electricity for the suburban electric trains between Euston, Broad Street, Watford and the Richmond line to Gunnersbury. On this duty they normally worked as a block train from Toton, indeed a precursor of today's block trains, and were rarely if ever to be seen elsewhere. They were coded LHZ, the Z indicating that they had a carrying capacity of 40 tons, capacity being 1794 cu ft, and tare weight 18 tons 15 cwt.

These vehicles were the largest to be found in the ordinary wagon diagram book, length over headstocks being 32ft 0in; the bogies had a wheelbase of 6ft 0in and were set at 18ft 0in centres, the wheels being 2ft 8½in diameter. These wagons were vacuum brake fitted and had handbrakes of a compound lever type.

Summary of Loco Coal Hoppers

Diag	Lot	Qty	Built	Year	Numbers
1668	149	50	Earlstown	1925	Various
2024	1199	450	Derby	1939	697000–697449

Above: Plate 21C is an ex-works picture in bauxite body colour of a loco coal wagon branded 'Crewe Works & Rhigos Colliery'. It was photographed on 8 February 1939.

Below: Plate 21D shows one of the special 40T bogie coal wagons, normally worked in block trains between the Toton area and the LMS power station at Stonebridge, which supplied power for the electric trains working between London and Watford.

The body was unusual in that it extended beyond the headstocks, being 33ft 6in long, while overall height was 10ft 9in; the body was also unusually wide being 8ft 11in overall. Four side discharge doors were fitted, each being no less than 15ft 4⅝in long. (Plate 21D).

Livery Details
Those vehicles repainted after 1936 would have received bauxite body colour with the insignia laid out in the standard positions as indicated in Chapter 2.

Summary of 40 Ton Coal Hopper

Diag	Lot	Qty	Built	Year	Numbers
1708	457	30	B'ham C&W	1929	189301–189330

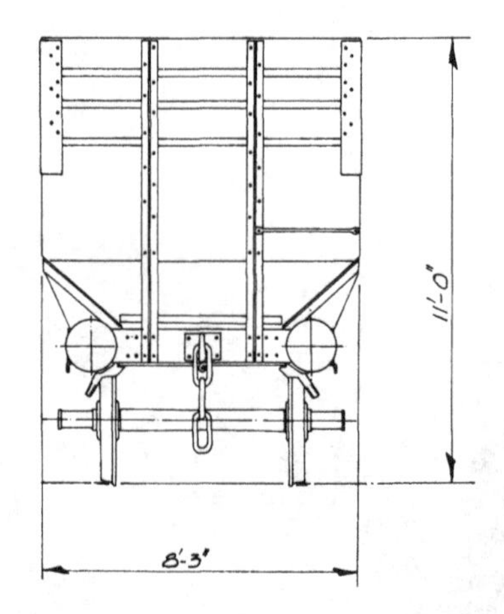

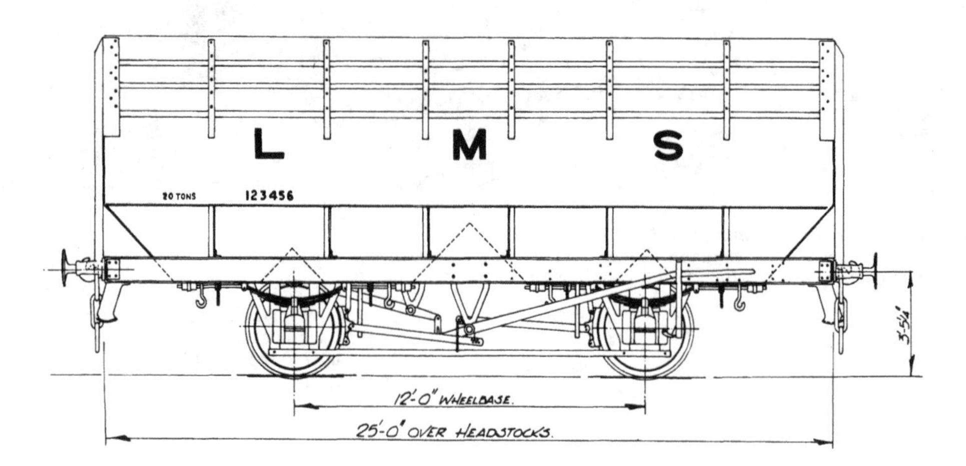

Coke Hopper
COKE is relatively light in relation to bulk, and this is reflected in the size of the vehicles to D1729 (fig 29), built between 1930–5. Their code CHO shows that they had a weight capacity of 20 tons, cubic capacity however was no less than 1380 cu ft, while the tare weight was 11 tons 5 cwts. (Plate 22A).

The vehicles were 25ft 0in long over headstocks on a 12ft 0in wheelbase, overall height being 11ft 0in. They were unfitted, the diagram stating that handbrakes of the double type were to be provided, though some at least had brakes of the 20 ton standard type. These vehicles were of the all-steel type, although the coke rails were made of timber, and had eight bottom discharge doors mounted in pairs on each side of the axles.

Livery Details
Plate 22A is the only known picture of this type of wagon but it is felt that the remarks applicable to the ballast hopper apply also to these vehicles.

Summary of Coke Hopper

Diag	Lot	Qty	Built	Year	Numbers
1729	552	100	B'ham C&W	1930	299900–299999
	879	50	Met Cam.	1935	699000–699049
	880	50	,, ,,	1935	699050–699099

Grain Hopper
THESE vehicles to D1689 (fig 30) were introduced in 1928, and building continued until 1949. They were of the covered type, and were unusual in being described as hopper grain vans, as opposed to wagons, their code BGV giving the alternative title of bulk grain van. (Plate 22B). Carrying capacity was 20 tons, cubic capacity being 1200 cu ft, with a tare weight of 10 tons 1 cwt. Length over headstocks was 21ft 6in on a 10ft 6in wheelbase, height to the top of the roof being 11ft 8⅞in, and overall height 12ft 8in.

Left: Fig 29 D1729 Coke hopper

Above: Plate 22A is an ex-works picture of a trade-built steel coke hopper.

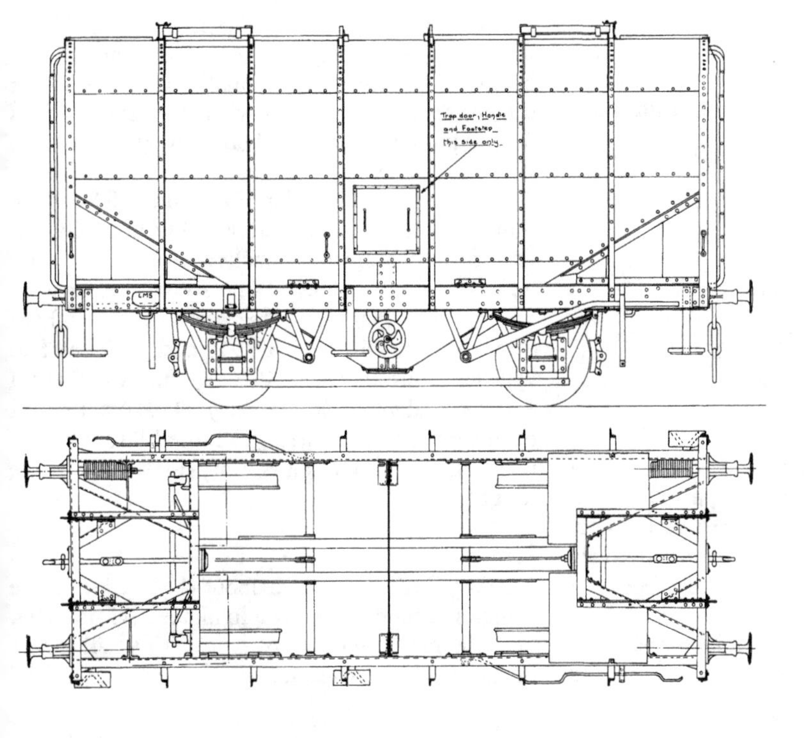

Left: Fig 30 D1689 Grain hopper van (End views, etc, page 92)

These vehicles were unfitted, and had a special type of independent handbrake at each end. Each set had a single short side lever attached to a cross shaft, supported by V-hangers on each side. From this cross shaft rodding went to the brake shoes, of which there was one on each wheel placed on the headstock side of the wheels. Depressing the lever thus pulled the shoes into contact, instead of the more usual arrangement whereby the brake shoes were pushed against the wheels. The levers were fitted diagonally opposite one another, to provide the either side requirement.

The roof was of three-arc form, and curved round to blend smoothly with the sides, while the angle iron stiffeners also ran up the sides and right over the roof. Two sliding doors were provided in the roof for filling purposes, while a single bottom opening 1ft 3in square, controlled by a handwheel on one side only, sufficed for discharging the contents. A trap door was provided in one side, to provide access to the interior for maintenance or repair when required, while to assist in inspection two small windows were provided at the top of each end; ladders diagonally opposite at each end, gave access to the top hatches.

Livery Details

First examples were lettered as 299022 but with the adoption of bauxite the legend BULK GRAIN moved down to the bottom of the side, in the same position, but smaller letters, with LMS, 20T, running number and N, all in the space between the first and second angle, the standard positions as described in Chapter 2. The tare weight and another N were at the extreme bottom right hand end of the vehicle.

Summary Grain Hopper

Diag	Lot	Qty	Built	Year	Numbers
1689	373	10	Derby	1928	
	402	15	,,	1928	
	723	10	,,	1933	Sample 299022
	755	15	,,	1934	
	833	10	,,	1934	701300–701309
	924	10	,,	1936	701310–701319
	1019	25	Hurst N'son	1937	701320–701344
	1208	10	Derby	1940	701345–701354

Ore Hopper

FIVE diagrams were issued by the LMS for these vehicles, all of which were very similar in size and appearance. Although all had a carrying capacity of 20 tons, being coded HPO, they were relatively small vehicles, as ores generally are very heavy in relation to bulk.

All were 17ft 0in long over headstocks, on a wheelbase of 10ft 6in; they were unfitted and had

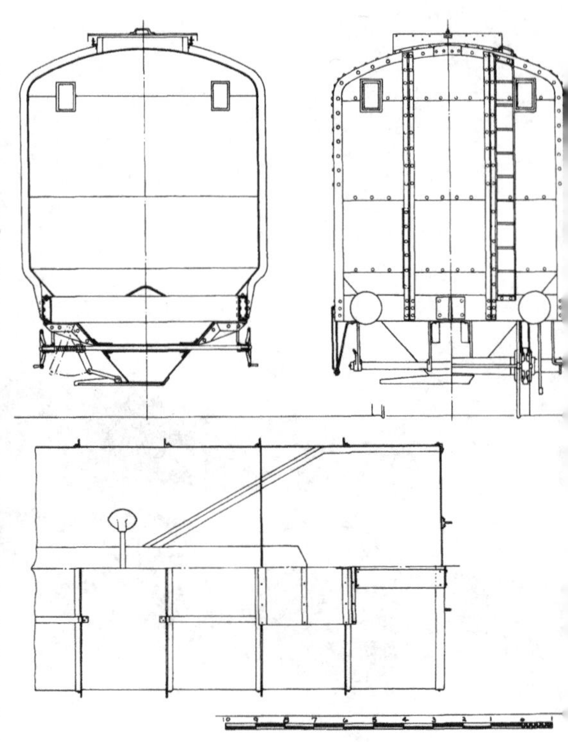

handbrakes of the double type. Two bottom discharge doors 9ft 6in long were provided, hinged along the longitudinal centre line.

Vehicles to the first diagram (D1669) were built between 1924–9, and were 7ft 9½in high overall, having a cubic capacity of 363 cu ft and a tare weight of 8 tons 5 cwt. (Plate 22C).

The remaining diagrams were for vehicles 8ft 6in high overall, having a cubic capacity of 430 cu ft; they varied fractionally in some other dimensions, the main reason for the various diagrams being that the bodies were built from various types of steel.

Diagrams D1893 (fig 31) and 1894 were dimensionally identical and were built in 1934, the first from Chromador steel, and having a tare weight of 7 tons 16 cwt. The second group were built from ordinary mild steel, with a tare weight of 8 tons 6 cwt. Diagrams D1941 and 1942 refer to vehicles built between 1936–8, and were again externally identical (Plate 22D). D1941 were built from copper bearing steel, and tared 8 tons 8 cwt, while D1942 does not specify material or tare weight.

Livery Details

Nothing further is known about the livery of these vehicles beyond that shown in plates 22C and 22D, but it is presumed that bauxite with small letters in the standard position was adopted after 1936.

Above: Plate 22B clearly shows the 1932 livery as used by the LMS for bulk grain hoppers.

Left: Fig 30A End view and plan view of body detail of grain hopper.

Below: Fig 31 D1893 Ore hopper wagon

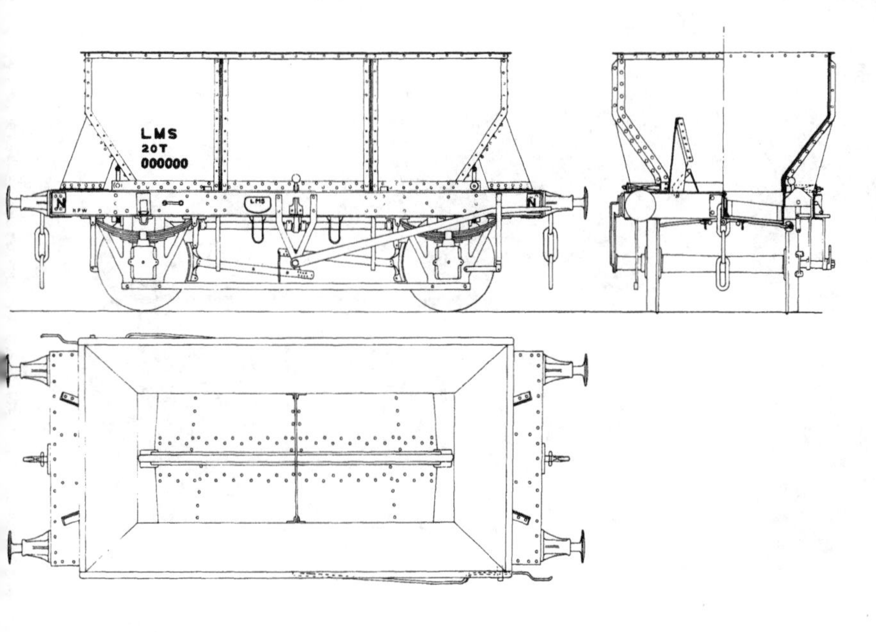

Plate 22C, *above*, is a 20T ore wagon built in 1924 by the Metropolitan Carriage, Wagon & Finance Company Limited. It is interesting because it shows the trade practice of displaying an advertisement board in the picture. Note the difference in size between this and the 20T coke hopper. A later version is shown in plate 22D, *below*, and the higher sides should be noted.

Summary of Ore Hoppers

Diag	*Lot*	*Qty*	*Built*	*Year*	*Numbers*
1669	74	50	Metcamell	1924	299051–299100
	75	50	B'ham C & W	1924	299101–299150
	76	50	Chas Roberts	1924	299151–299200
	77	50	Cammell Laird	1924	299201–299250
	465	150	Met Cammell	1929	189571–189720
	466	50	G R Turner	1929	189721–189770
	467	100	Hurst Nelson	1929	189771–189870
	468	100	Craven & Co	1929	189871–189970
1893	789	50	Met Cammell	1934	690000–690049
1894	790	50	Met Cammell	1934	690050–690099
	866	200	" "	1934	690100–690299
	867	50	Hurst Nelson	1934	690300–690349
1941	940	350	Met Cammell	1936	690350–690699
	941	200	B'ham C & W	1936	690700–690899
	966	100	" "	1936	691000–691099
	967	150	Met Cammell	1936	691100–691249
	1020	150	B'ham C & W	1937	691250–691399
	1133	300	Met Cammell	1938	691400–691699
	1134	200	B'ham C & W	1938	691700–691899
1942	942	100	Gloucester	1936	690900–690999

Lime Hopper

THESE vehicles were converted from ore hoppers, of diagrams D1893, 1894 and 1941. They were given diagram number D2194, and were converted without the benefit of a lot number. The conversion involved fitting a sloping roof, with a loading hatch in each side.

Zinc Oxide Hopper

THE vehicles to D2139 were also conversions, the basic vehicle being an open hopper built by the Glasgow & South Western Railway. They were 19ft 0in long over headstocks on a 10ft 6in wheelbase, and had a carrying capacity of 20 tons; capacity was 524 cu ft, and tare weight 10 tons. They were unfitted and were provided with screw down hand brakes. The conversion again involved fitting a roof, with two loading hatches, and short ladders on each side for access purposes. Overall height was 9ft 0¾in, and the diagram states that 12 vehicles were so treated in 1948.

Soda Ash Hopper

THESE vehicles to D2128 were of the covered type built in 1948. They were 20ft 0in long over headstocks on a 10ft 6in wheelbase; carrying capacity was 20 tons, tare weight being 10 tons 12 cwt, while overall height was 10ft 7½in. They were unfitted and were provided with an unusual form of Morton hand brake; this operated on two wheels on one side only, the brake shoes being on the headstock side of the wheels. Four large hatches were provided in the roof, with no fewer than eight small bottom discharge doors.

Livery Details

These vehicles were built to an LMS diagram in 1948, and were finished in a dark body colour (LMS bauxite) with the number M689300 beneath 20T located half way up the side at the left hand end, with SODA ASH, TO WORK BETWEEN WINNINGTON LMR (CLC) AND ST. HELENS LMR to the right between the first and second angle irons.

Summary of Soda Ash Hoppers

Diag	*Lot*	*Qty*	*Built*	*Year*	*Numbers*	*Remarks*
2128	1469	6	Derby	1948	292089–292094	For LNER
		15			689300–689314	For LMS

CHAPTER 12

FOUR WHEELED SPECIAL WAGONS

VEHICLES classed by the LMS as special wagons had their own diagram book. Unlike the ordinary wagon stock, where standard LMS vehicles were kept apart from pre-grouping wagons, special wagons were combined into one volume. The vehicles described in this and the following chapter are those built to LMS lots, although one or two are to pre-grouping designs.

Special wagons were not given diagram numbers, and are identified by diagram page number only. Another difference was that many diagrams consisted of two pages, the first showing the diagram, while the second recorded running numbers and detailed information on running restrictions.

Four-wheeled vehicles were relatively few, and the reader may well wonder why they fell into the special wagon category, since some vehicles already described were just as specialised in their duties. Unfortunately there is no clear cut answer and it can only be put down to the vagaries of the system.

Codes

As already mentioned some wagon codes consisted of three letters, the third of which indicated the carrying capacity. This practice was more widespread among special wagons, and to prevent repetition the letters used, together with the carrying capacity they represented, are given in the list below. It can be assumed that they apply to all vehicles in this and the following chapter, unless stated otherwise.

B = 5 tons	N = 18 tons	U = 52 tons
D = 6	O = 20	Y = 60
F = 8	R = 25	C = 65
H = 10	S = 30	T = 70
K = 12	A = 35	Q = 80
T = 14	I = 36	X = 100
M = 15	Z = 40	L = 160
J = 16	P = 50	

Implement Trucks

THESE were fairly long vehicles, used for the transport of farm implements and similar types of machinery. They were of all-steel construction, the majority having a timber floor or decking. The centre portion of the vehicle between the wheels was depressed, being about 2ft above rail level, while the ends sloped upwards, thus allowing for normal buffers.

The first batch, diagram P52, were built between 1925–9, 27ft 0in long over headstocks on a 21ft 0in wheelbase. The depressed portion was 14ft 0in long, while height to the top of the headstocks was 3ft 10in. This particular diagram was sub-titled 'chemical pan', and could be fitted with timber beams 12in square, which ran the full length of the vehicle supported at the ends by cross baulks 15in high by 9in thick; a well 7ft long by 2ft 0in wide was thus provided in the centre of the floor. They were unfitted and provided with handbrakes operated by very short levers. Carrying capacity was 20 tons, their code being UIO, while the tare was 10 tons 17 cwt with, or 8 tons 17 cwt without, the timber beams.

The next two diagrams were for vehicles 34ft 0in long over headstocks, on a 25ft 0in wheelbase, having wheels 2ft 8½in diameter, the depressed portion of the body being 16ft 0in. Carrying capacity was again 20 tons, and the code U10.

P52A referred to wagons built between 1931–7; they were unfitted, with handbrakes of the short side lever type, and tare weight 10 tons. (Plate 23A). P52B wagons differed only in being fully fitted; they were built in 1940 and had a tare weight of 10 tons 16 cwt.

The final diagram in this group, P54A, was for wagons built in 1944; they were 30ft 0in long over headstocks on a 22ft 0in wheelbase, the depressed

Top right: Plate 23A shows an Implement truck as built in ex-works livery.

Right: Fig 32 P59A Traction truck

Summary of Implement Trucks

Diag	Lot	Qty	Built	Year	Numbers
P52	150	11	Derby	1925	18866, 19890, 23388, 30554
	354	15	"	1928	30564, 34797, 50477, 77988
	429	5	"	1929	80675, 96080, 117588. 178614, 178841, 179315, 189986–189990, 294262, 297160, 297162, 300009, 300017, 300020, 300022, 300026, 300029, 300034, 300082.
P52A	606	5	Derby	1931	299785–299789
	607	5	"	1931	299790–299794
	1018	10	R & W Maclellan	1937	700600–700609
P52B	1212	10	Wolverton	1940	700610–700619
P54A	1342	30	LNER	1944	700700–700729

portion of the body being 16ft 0in long. They had a carrying capacity of 25 tons, the code being UIR, and were fully fitted with short side lever hand-brakes; tare weight was 13 tons 12 cwt.

Livery Details

The absence of body side left little opportunity for variation and generally with grey livery the LMS was centrally placed, and in bauxite it was to the left hand end with often the number only on the number-plate.

Traction Trucks

THESE vehicles were very similar in appearance to the implement wagons, but were much smaller, carrying capacity in all cases being 12 tons. They were coded TRK, and all were 20ft 0in long over headstocks on an 11ft 0in wheelbase: the length of the flat portion of the body was 14ft 0in.

The first diagram (P59), was of Midland origin dating back to 1914, the LMS lot being built in 1929. The diagram shows what appears to be an all-wood vehicle with a practically level top face,

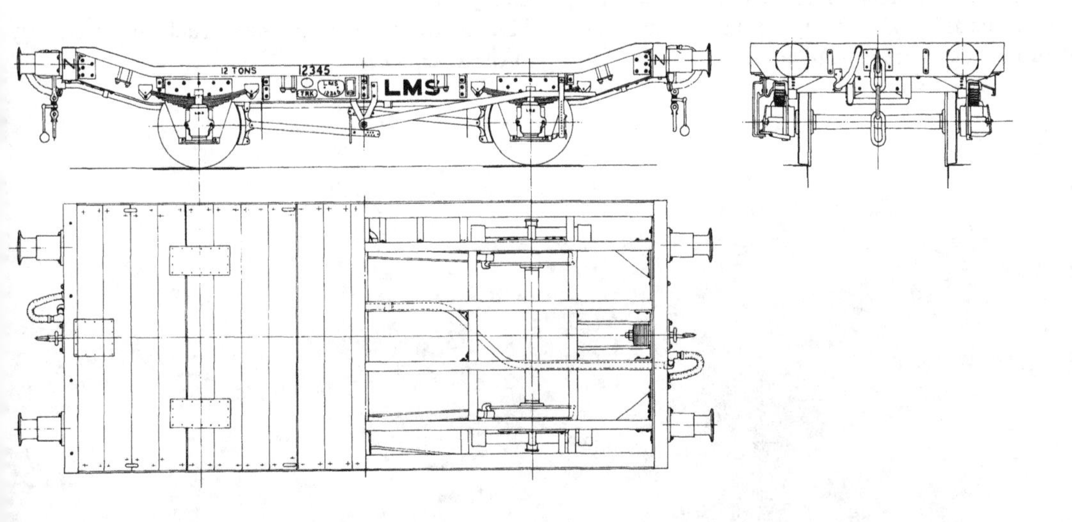

Above: Plate 23B Traction truck No 700426 was photographed on 12 March 1940; note the wheel chocks nailed to the floor & method of roping the load.

Below: Plate 23C Deep case wagon, developed from MR design, in original livery style. A. E. West

height to the centre level portion being 3ft 1in. Photographs in our possesion, which include one of the original MR vehicles, show them to have been almost identical to the later LMS diagrams. The description which follows can therefore be taken as applying to all these vehicles.

The LMS diagrams give a height of 3ft 1in, to the top of the level portion of the floor, the sloping ends being 9½in above this at the headstocks. The solebars were of steel channel, with the flat face outside, the axleguards being on the outside, while the headstocks were of timber.

Wheels were 2ft 8½in diameter. Those to diagram P59A (fig 32), built between 1931–7 being unfitted, though some were piped (plate 23B). Tare weight was 6 tons 8 cwt, and normal long lever double handbrakes were provided. Those to diagram P59B, built in 1939 were identical, apart from being fully fitted; tare weight was 7 tons 3 cwt.

Livery Details

The remarks about implement trucks apply equally to these vehicles.

Summary of Traction Trucks

Diag	Lot	Qty	Built	Year	Numbers
P59	408	40	Wolverton	1929	189431–189470
P59A	604	10	Derby	1931	299856–299865
	605	10	,,	1931	299866–299875
	827	20	,,	1935	700400–700419
	925	10	,,	1936	700420–700429
	1028	25	,,	1937	700430–700454
P59B	1202	10	Derby	1939	700455–700464

Deep Case Trucks

THESE are extremely difficult vehicles to describe in words, and the reader is referred to plate 23C, suffice it to say here that they resembled a small girder bridge on wheels. The main girders were very close to rail level, and this allowed tall light loads to be carried. The vehicles to diagram P66A were built in 1928; they were 40ft 0in long over headstocks on a 20ft 0in wheelbase. Carrying capacity was 12 tons, the code being DCK. Not surprisingly they were unfitted and were provided with side lever handbrakes, tare weight being 9 tons 9 cwt. Some further vehicles of similar type were the ex MR skeleton wagons. Several of these vehicles were partly renewed to LMS lots, and the known details are included in the summary table.

Livery Details

Plate 23C is the only known photograph and in bauxite livery a similar style would have been adopted.

Summary Diagram P66A of Deep Case Trucks

Lot	Qty	Built	Year	Numbers
367	4	Pickering	1928	117288–117291

Ex MR skeleton wagons

Lot	Details	Year	Number
10	10 ton cap'y renewed Derby	1924?	117297
86	,, ,, ,, ,, ,,	1924?	117293
104	,, ,, ,, ,, ,,	1925?	117295

Glass Trucks

THESE vehicles were used to carry large sheets of plate glass, the LMS vehicles to diagram P75 being 26ft 0in long over headstocks on a 21ft 0in wheelbase. They consisted of a steel underframe, with steel well lined at the bottom with timber; the well length was 20ft 0in, and width 4ft 1⅞in, the floor being 11in above rail level.

A large angle-iron frame was secured to each solebar, which gave an overall height of 11ft 7¾in, and to this were fastened screw jacks with which the cases holding the glass could be clamped securely. Carrying capacity was 12 tons, the code being GLK, while the tare weight was 8 tons 3 cwt; they were unfitted and had handbrakes of the double type.

Livery Details

Plate 23D is typical of the livery style used with these vehicles. The change to bauxite resulted in the tonnage, company ownership and running number being compressed against and to the right of the numberplate and code plate, with the tare weight at the extreme right hand end of the vehicle.

Trollies

ALL trolley wagons were of the well type, the load carrying portion of the vehicle being very close to rail level. The ends of the well were vertical, and raised to sufficient height to contain the wheels, buffers and drawgear.

The first diagram (P85A) was for vehicles built in 1928 sub-titled 'chemical pan'. They were 21ft 0in long over head-stocks on a 16ft 0in wheelbase, and had a well length of 12ft 0in, the top of the well being 1ft 2½in above rail level. Timber baulks 12in square by about 18ft long were provided, resting on transverse baulks 15in high by 9in thick. Carrying capacity was 12 tons, the code

Summary Diagram P75 Glass Trucks

Lot	Qty	Built	Year	Numbers
224	2	Derby	1926	200064, 202918, 206320, 206935, 212187
353	8	,,	1928	221096, 258391, 268124, 268127, 355323
409	30	Wolverton	1928	168876–168905
1017	10	G R Turner	1937	700800–700809
1286	10	Derby	1940	700810–700819
1423	6	,,	1947	700820–700825

Above: Plate 23D Glass wagon in early LMS livery, photographed in 1926.

Below: Fig 33 P96B 20T trolley

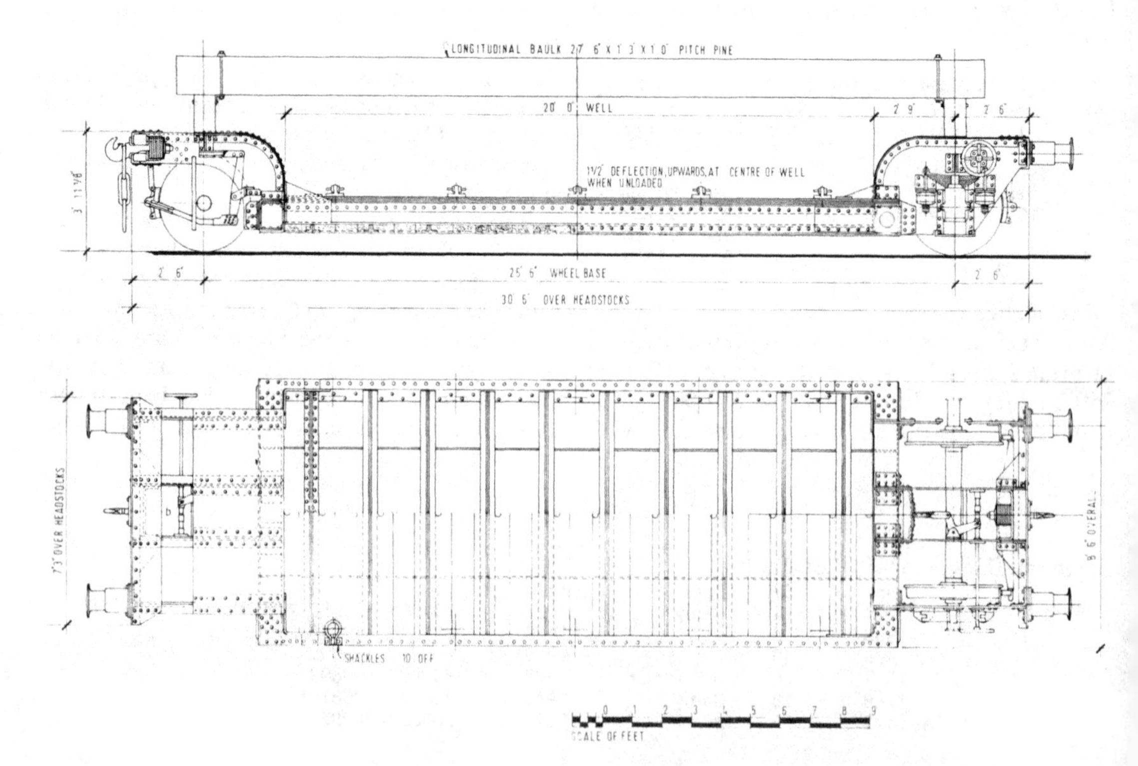

Summary of Trollies

Diag	*Lot*	*Qty*	*Built*	*Year*	*Numbers*
P85A	352	3	Derby	1928	233925, 233926, 266112
P96A	417	9	Derby	1929	168964–168972
	828	5	"	1935	700000–700004
	1213	4	Wolverton	1939	700010–700013
P96B	608	6	Wolverton	1930	299876–299881
	829	5	Derby	1935	700005–700009
P106	223	6	B'ham C & W	1925	117577, 117578, 117580 117582, 117587, 117591

being TYK, while tare weight was 9 tons 4 cwt with, or 7 tons 17 cwt without, the baulks. These vehicles were unfitted, and had handbrakes with very short levers.

Vehicles to the second diagram (P96A) were built between 1929–39; they were 29ft 0in long over headstocks on a 24ft 0in wheelbase, and had a well length of 20ft 0in, height to top of well being 1ft 5in. Timber baulks were again provided, in this case being 12in by 15in in section and about 26ft long. Carrying capacity was 20 tons, the code being TYO; tare weight was 11 tons 15 cwt with, or 10 tons 14 cwt without, the baulks. The wagons were unfitted, with short lever handbrakes.

The next diagram P96B (fig 33) applied to vehicles built between 1930–5. They were 30ft 6in long over headstocks on a 25ft 6in wheelbase. All other dimensions were as for P96A, tare weight in this case being 14 tons with, or 12 tons 10 cwt without, baulks. The final diagram in this group was P106, for wagons built in 1925; they had a carrying capacity of 25 tons, the code being TYR. Length over headstocks was 27ft 0in on a 22ft 0in wheelbase, well length being 18ft 0in. The wheels were 2ft $8\frac{1}{2}$in diameter, and the well had a wooden floor which could be removed in sections when required. Wooden baulks were again provided, being 12in square by about 24ft 0in long; tare weight was 10 tons 16 cwt with or 9 tons 12 cwt without the baulk and the wagons were unfitted. (See plate 25A).

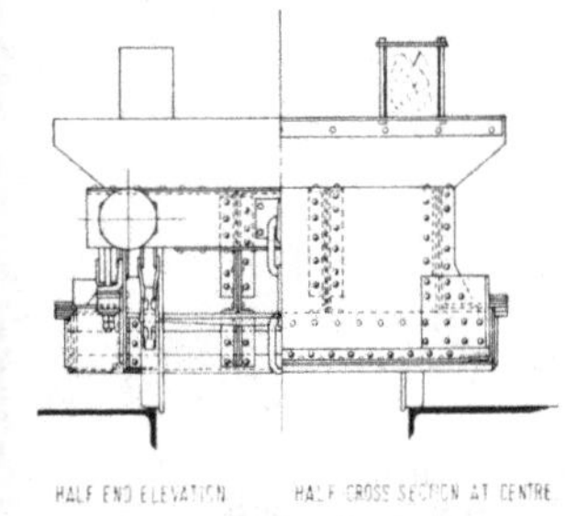

Livery Details

Like implement trucks, a similar style was followed; with P96B the deeper side resulted in a rather peculiar shaped S which was rather flattened, because of the width adopted for the letter, but lack of height available.

Chaired Sleeper Trolley

LITTLE is known of the dimensions of these vehicles, as they were not allocated a diagram number. They were similar in appearance to the other four-wheeled trollies, but had metal end walls to the raised portion, and stanchions fitted to the side of the well, carrying capacity being 20 tons. (See frontispiece).

Summary of Chaired Sleeper Trollies

Lot	*Qty*	*Built*	*Year*	*Numbers*
365	1	Derby	1928?	
423	44	"	1928?	168926

Match Wagons to Armour Plate Trucks

THIS diagram illustrates the difficulty the railways sometimes had in classifying vehicles, for although the vehicles were branded 'Not to be loaded. For use as check wagons only', and were used as their title suggests with the armour plate wagons described in the following chapter, they were given diagram number D2003 in the ordinary wagon book. They consisted of a steel underframe of 12in deep section, 18ft 2in long over headstocks on a 10ft 0in wheelbase. On top was a $\frac{3}{8}$in thick steel deck, and they were probably ballasted, as the tare weight was 13 tons 1 cwt. Little else is known of them since no lot numbers appear to have been given; a note on P2A states that two such vehicles were kept at Sheffield, their numbers being 116347, 116348.

CHAPTER 13

BOGIE SPECIAL WAGONS

IN this chapter we examine the remaining special wagons, which include some of the heaviest vehicles ever to run on British railways, and of whose classification there can be no doubt whatsoever.

Armour Plate Trucks

THESE vehicles, as their title suggests, were used to carry very heavy plates. They were flat wagons built from heavy steel sections, in most cases being open topped, that is without a floor of any kind.

The first diagram P2 was actually a Midland design, built between 1892 and 1898. They were given an LMS lot number, however, and this possibly covered a partial renewal, as the diagram quotes a carrying capacity of 40 tons, while the lot book quotes 45 tons. The code was AZ, correct for the former tonnage; the lot book may therefore be in error here. Length over headstocks was 24ft 0in, the bogies of 6ft 0in wheelbase spaced at 12ft 0in centres. These vehicles had a tare weight of 18 tons 9 cwt; they were unfitted, being provided with lever type hand brakes.

Although space precludes inclusion of full route restriction details which applied to all vehicles in this chapter, those quoted here may be regarded as typical. The minimum curve that the wagon could negotiate was one chain. The load was not to exceed 30 tons, equally distributed, when loaded to Furness section stations, or 22 tons when loaded to Trafford Park and the Manchester Ship Canal lines. The Southern Railway was to be consulted before loading to the SEC section, while if two of these wagons were travelling fully loaded in the same train, two comparatively light wagons had to be marshalled between them.

The next diagram P2A was for a vehicle built in 1937, rated to carry 100 tons, the code being HAX. The body was formed of heavy H-section girders, 26ft 0in long over headstocks. Heavy duty bogies having a wheelbase of 5ft 0in were fitted at 11ft 6in centres, and this vehicle had a tare weight of 18 tons 19 cwt, or if fitted with a 3in thick armour plate top tare was 28 tons 14 cwt. It was stated on the diagram that this vehicle was for internal working only, and this may explain the apparent complete lack of brakes, while a further note stated that match trucks for this vehicle were kept at Sheffield; these might have been the vehicles to D2003 described in the previous chapter.

The vehicles to diagram P2B were built in 1939 and had a carrying capacity of 40 tons, code being AZ. They were 24ft 0in long over headstocks, with bogies of 6ft 0in wheelbase set at 12ft 0in centres. They had 2ft 8½in diameter wheels, and screw down handbrakes, tare weight being 13 tons 15 cwt. The final vehicles in this group, diagram P3A, were built in 1937–8 and had a carrying capacity of 55 tons, code being AE. They were 30ft 0in long over headstocks, with 8ft 0in wheelbase bogies set at 16ft 0in centres. They had screw down handbrakes, and had a wooden floor or decking, tare weight being 16 tons 3 cwt.

Livery Details

Plate 24A indicates a typical layout with bauxite body colour; in grey the letters LMS would probably have been larger and centrally placed upon the vehicle.

Bogie Bolster Wagons

THESE vehicles were among the most numerous of the special wagons, and this, plus the fact that they were used over a wide area will have made them familiar to many readers. The LMS issued several diagrams to cover these vehicles, which fell

Summary of Armour Plate Trucks

Diag	*Lot*	*Qty*	*Built*	*Year*	*Numbers*
P2	243	12	Cohen & Armstrong Renewed 1926	1882–98	116065–116074 41623–41626, 11019, 34841, 41628
These numbers are those quoted in the diagram book, but not all can apply to this lot.					
P2A	1061	1	Derby	1937	700200
P2B	1214	10	Wolverton	1939	700220–700229
P3A	1065	6	Wolverton	1937	700201–700203 (3LMS)
	1144	1	,,	1938	217329–217331 (4LNER) 217295

Above: Plate 24A Armour plate truck, one of seven built in 1937/8, four for the LNER, three for the LMS.

Below: Plate 24B A 'Warwell' converted by BR into a bogie bolster wagon.

into two main groups, the first being conventional flat topped vehicles, while the second group consisted of ex-Ministry of Supply well wagons (warwells), modified for use as bolsters. The vehicles were of several different carrying capacities, which thus forms a convenient means of sub-division.

30 Ton Capacity

These vehicles were of both conventional and warwell type, the conventional examples being coded BBS. They were 45ft 0in long over headstocks, and had diamond framed bogies of 5ft 6in wheelbase set at 34ft 6in centres. The main frames were of 9in deep sections with angle iron trusses, and the body was completed by a wooden floor with sides and ends of timber 8in high. Four bolsters were provided at 11ft 6in centres, the outer pair being over the bogie centre lines.

The first diagram appeared in both the ordinary and special wagon diagram books, as D1682 and P11 respectively, and was an MR design; a total of 407 vehicles were built, of which 277 were to LMS lots. These vehicles had a tare weight of 14 tons 11 cwt, and the LMS examples were built between 1926–8.

Two further diagrams were issued, P11A (fig 34) for wagons built in 1936 and P11C for those built in 1939, the tare weight being 15 tons 5 cwt and 15 tons 9 cwt respectively. This apart the three diagrams were virtually identical; all the vehicles were unfitted and had short lever type handbrakes.

Two diagrams were also issued to cover warwell conversions, both of which were derived from the same basic vehicle. They were 43ft 0in long over headstocks, and had diamond framed bogies of 5ft 9in wheelbase with 2ft 9in diameter wheels, set at 33ft 0in centres. The first diagram (P5B) showed six bolsters set at 7ft 6in centres; the middle pair were set in the well and had plate steel supports beneath them to raise them to the same level as the remaining bolsters, tare weight being 29 tons 5 cwt. (Plate 24B). The second diagram P11D was for bogie rail wagons, and in

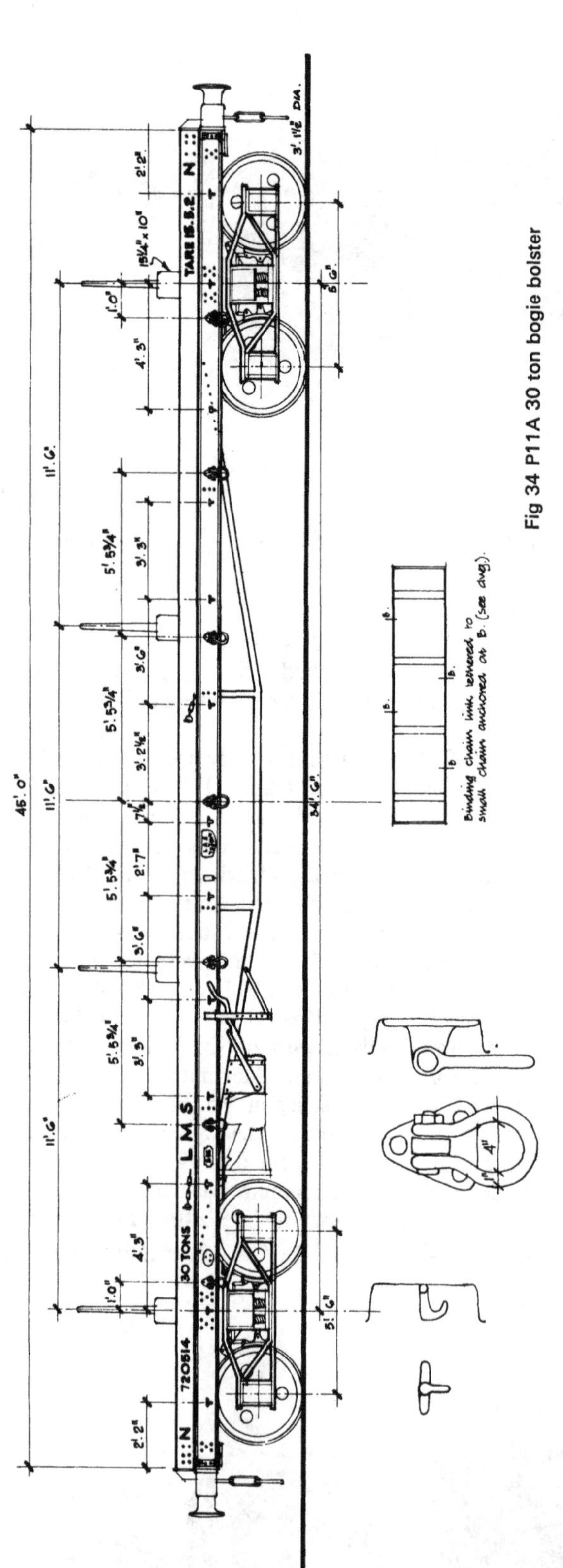

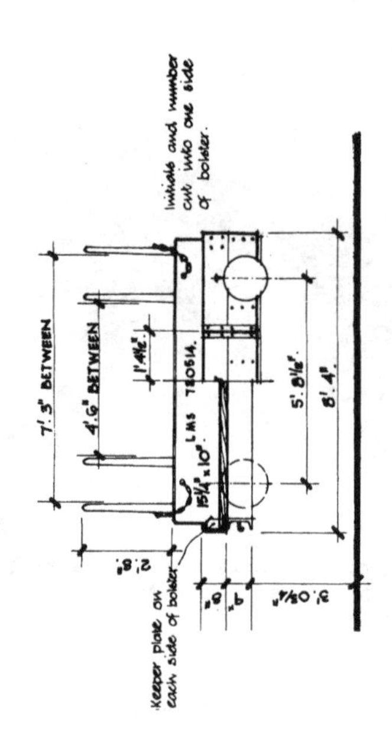

Fig 34 P11A 30 ton bogie bolster

this case a timber deck was carried across the well supported by light channel section girders. Four bolsters were fitted, the centre pair at 9ft 11¼in centres being over the well with fabricated steel supports between them and the well deck; the outer bolsters were at 11ft 6⅜in centres from the centre pair. Tare weight was 28 tons 17 cwt, and all of these conversions were vacuum fitted and provided with screw down handbrakes; diagram P5B was coded Bolster B, while P11D were coded BBS. All these conversions were carried out in 1947–9.

35 Ton Capacity

Two diagrams were issued for vehicles of this capacity, their code being BBA. They shared a common length of 52ft 0in over headstocks, with frames of 10in deep sections trussed with angle iron. Bogies were of the diamond framed type, with 2ft 8½in diameter wheels, and set at 40ft 0in centres.

Vehicles to the first diagram P15A were built in 1938, and had bogies of 5ft 6in wheelbase, the body being formed from two planks with a total height of 14½in, while the ends were arranged to drop down. No bolsters appear to have been fitted to these vehicles, which tared from 18 tons 6 cwt to 18 tons 16 cwt.

Vehicles to the second diagram (P15B) were built in 1940, and had bogies with a 6ft 0in wheelbase. The body in this case was somewhat shallower, with sides about 7in high and fixed ends; five bolsters were provided, one on the centre line and the others at 12ft 8½in and 12ft 8⅞in centres.

Summary of 30 Ton Bogie Bolster

Diag	Lot	Qty	Built	Year	Numbers
P11	287	89	Wolverton	1926	
	348	50	Metro C & W	1927	
	349	50	Gloucester	1927	
	392	88	Wolverton	1928	
LMS built vehicles carried various numbers including the block 290017–290082					
P11A	937	50	Hurst Nelson	1936	720500–720549
	983	25	" "	1936	720550–720574
P11C	1210	30	Wolverton	1939	720575–720604
P5B	1547	38	Wolverton	1949	721200–721237
P11D	1497	50	D'by & W'ton	1947/8	748300–748349

Summary of 35 Ton Bogie Bolster

Diag	Lot	Qty	Built	Year	Numbers
P15A	1162	10	Hurst Nelson	1938	720700–720709
P15B	1211	20	Wolverton	1940	720710–720729

Tare weight was 18 tons 13 cwt; both diagrams were unfitted and had lever type handbrakes mounted on the bogies. (Plate 24C).

40 Ton Capacity

These vehicles to diagram P18A were built in 1935, and coded BBZ. Dimensionally they were identical to the 35 ton capacity vehicles to diagram P15B described above, the only difference appearing to be that the lever handbrake was attached to the underframe; tare weight was 18 tons 5 cwt.

Summary of 40 Ton Bogie Bolster P18A

Lot	Qty	Built	Year	Numbers
841	20	Wolverton	1935	720000–720019

42 Ton Capacity

Two diagrams were issued for this capacity, and no code appears to have been allocated. All the vehicles were 52ft 0in long over headstocks, with bogies at 40ft 0in centres, and had trussed channel underframes. The bogies were again of the diamond frame type, and were fitted with 2ft 9in diameter wheels.

The first lot (diagram P18B) were built in 1947, and had bogies with 5ft 6in wheelbase. They were classed as bogie plate wagons, with drop sides in two sections and fixed ends, the height of the body being about 12in; tare weight was 18 tons 17 cwt. The second lot (diagram P18C) were built in 1948, and had bogies with a 6ft 0in wheelbase. These vehicles were also almost identical to the 35 and 40 ton vehicles to P15B and P18A with five bolsters at the same spacing as before, tare weight in this case being 18 tons 12 cwt.

Summary of 42T Bogie Bolster

Diag	Lot	Qty	Built	Year	Numbers
P18B	1422	50	Derby	1947	721500–721549
P18C	1513	66	Derby	1948	720020–720085

50 Ton Capacity

Seven diagrams were issued for the vehicles of this capacity; they were coded BBP, and were

Below: Plate 24C A 35 ton bogie bolster as built, and finished in bauxite livery; one of 20 turned out at Wolverton in 1940.

Plate 24D A 50T bogie bolster wagon built at Wolverton in 1941.

used mainly for the carriage of rails. All were 62ft 0in long over headstocks, with trussed channel underframes, and had channel framed bogies of carriage type with a wheelbase of 8ft 0in, the wheels being 2ft 9in diameter. Handbrakes of the short lever bogie-mounted type were provided in all cases. These vehicles had a wooden floor or decking without sides, but with hinged flaps across the ends, which were folded back on top of the floor when not in use but which could be swung over partially to cover the buffers when required.

The diagrams fall into two groups, of which the first showed the bogies set at 48ft 0in centres; five bolsters were provided, one on the centre line with the remainder at 12ft 0in centres, the only difference between the diagrams being the height from rail to top of floor and bolsters as summarised below:

Diag	*Built*	*Tare*	*Height to Floor*	*Height to Bolster*
P19A	1929	25T 5c	4ft $2\frac{1}{4}$in	5ft $0\frac{5}{8}$in
P19B	1936/8	24T 10c	4ft $2\frac{5}{8}$in	5ft $3\frac{1}{2}$in
P19C	1937/9	23T 10c	4ft $2\frac{1}{2}$in	4ft $6\frac{1}{2}$in
P19D	1939	23T 5c	4ft $2\frac{1}{2}$in	4ft $2\frac{7}{8}$in

The second group was practically identical in appearance, but had the bogies set at 45ft 0in centres, with five bolsters set at 11ft 3in centres. Again heights to floor and bolster were the only distinguishing features and these are again summarised below. (Plate 24D).

Diag	*Built*	*Tare*	*Height to Floor*	*Height to Bolster*
P19E	1940	23T 2c	4ft $2\frac{3}{8}$in	4ft $2\frac{3}{4}$in
P19F	1941	24T 18c	4ft $2\frac{1}{2}$in	5ft $3\frac{3}{8}$in
P19G	1942	23T 18c	4ft $2\frac{3}{8}$in	4ft $6\frac{3}{8}$in

Summary of 50 Ton Bogie Bolster

Diag	Lot	Qty	Built	Year	Numbers
P19A	420	12	Derby	1929	168908–169919
P19B	988	8	Fairfield	1936/7	720900–720907
	1138	30	,,	1938	720908–720937
P19C	1080	2	Wolverton	1937	748000–748001
	1146	13	,,	1938/9	748002–748014
P19D	1159	25	Wolverton	1939	748015–748039
P19E	1286	25	Wolverton	1940	748040–748064
	1292	25	,,		721025–721049
P19F	1293	25	,,	1941	720938–720962
P19G	1319	25	Wolverton	1942	748065–748089

Livery Details
With the conventional flat topped vehicles plate 24D shows the conventional later style of livery detail. When in grey several variations were used but the one generally adopted was to have LMS well spread out with the running number at each end of the vehicle and the tare weight on the sole-bar at the left hand end of the vehicle. A point worth noting was the practice of cutting the wagon number into the bolsters.

The warwells came into service during the bauxite livery period and therefore conformed to the style of having the running number and company ownership at the left hand end with the tare weight at the right hand end. Often various legends appeared in the centre of the vehicle, 'stores dept' loading instructions etc.

Trollies
THE collective name 'trolley' covered a variety of vehicles which ranged from the little four-wheeled examples described in chapter twelve, to the enormous multi-bogied trolley with a carrying capacity of 120 tons. These vehicles have been divided into groups according to capacity.

20 Ton Capacity
Two diagrams were issued for this capacity, coded BTO. The first diagram P97A was for vehicles built between 1929–40; they were unusual in that the bogies extended beyond the main body which was of the well type. The body, fabricated from steel plate and channel sections, was 53ft 8½in long overall with a 40ft 0in long well; timber baulks the same length as the body were provided, supported at each end on cross baulks. Channel framed bogies with a 5ft 6in wheelbase were fitted at 52ft 6in centres; length over bogie headstocks was 61ft 6in. The wheels appear to have been 2ft 9in diameter, and the headstocks were secured to the top of the bogie frames in order to maintain the buffers at the correct height. Tare weight was 27 tons 15 cwt with, or 25 tons 7 cwt without, the timber baulks, and handbrakes of the short lever bogie mounted type were provided.

The second diagram P98A was for vehicles built in 1940. They were conventional well vehicles 42ft 0in long over headstocks, with a well 20ft 0in long. Bogies of 5ft 6in wheelbase with 2ft 8½in diameter wheels were fitted at 32ft 6in centres. Timber baulks 15in by 12in by approximately 27ft long were supported on cross baulks placed near to the well ends. Tare weight was 19 tons 14 cwt with, or 17 tons 10 cwt without, the baulks, and screw down handbrakes were provided.

Summary of 20 Ton Trolley

Diag	Lot	Qty	Built	Year	Numbers
P97A	421	6	Derby	1929	189980–189985
	1131	3	Fairfield	1938	700370–700372
	1269	8	Derby	1940	700377–700384
P98A	1203	4	Derby	1940	700373–700376

35 Ton Capacity
Only one diagram was issued for this capacity, the vehicles to P118 being built in 1928. They were conventional well wagons fitted with large trestles in order to carry large plates on edge. Length over headstocks was 56ft 0in with a well length of 35ft 0in; bogies of 5ft 6in wheelbase having 2ft 8½in diameter wheels were set at 46ft 6in centres.

Summary of 35 Ton Trolley

Diag	Lot	Qty	Built	Year	Numbers
P118	366	3	G. R. Turner	1928	218857, 249992,

40 Ton Capacity
The LMS issued nine diagrams for vehicles of this capacity; they were coded BTZ apart from the first two diagrams, which appear to have been coded Trestrol MD. All the vehicles in this group were of conventional well type.

The first two diagrams applied to vehicles fitted with trestles, to enable them to carry large plates. The first to diagram P122A were built in 1930 and were 61ft 9in long over headstocks with a 40ft 3in well. They had 5ft 6in wheelbase bogies with 2ft 8½in diameter wheels at 52ft 3in centres. The bogies were of the diamond framed type and had short lever hand brakes mounted on them; tare weight was 26 tons 12 cwts with, or 24 tons 15 cwt without, trestles. (Plate 25B).

The second diagram P122B was for vehicles built in 1937–9; they were 55ft 6in long over headstocks with a 34ft 0in well. The bogies were the same as for P122A and were set at 46ft 0in centres. Tare weight was 26 tons 19 cwt with, or 24 tons 7 cwt without, the trestles.

Above: Plate 25A A 20T well trolley built in 1935; it could be used with or without the timber baulks.

Below: Plate 25B A 40T bogie trolley with trestles. The tare weight with trestles was 26 ton 12 cwt, and without 24 ton 15 cwt. Four were built in 1930.

Bottom: Plate 25C A 40T bogie trolley built in 1925. Note the open well and end construction. The baulks could be removed should circumstances so dictate.

Summary of 40 Ton Trollies

Diag	Lot	Qty	Built	Year	Numbers	
P122A	462	4	Derby	1930	189991–189994	
P122B	1064	15	Derby	1937	LMS 700300–700307	14 to LMS
	1145	2	Wolverton	1938	700312–700317	10 to LNER
	1165	12	Derby	1939	LNE 203889–203896	6 LNER, No.
					217296–217297	not known
P126	220	8	B'ham C & W	1925	117556, 117557, 117559, 117562, 117564, 117568, 117571.	
P126A	609	6	Wolverton	1930	299882–299887	
P126B	1132	4	G R Turner	1938	700308–700311	
P126C	1276	4	Wolverton	1940	700318–700321	
P129	228	8	Craven	1926	200185, 202981, 203553, 208381, 211061, 227830, 245739, 245740	
P129A	610	6	Wolverton	1930	299888–299893	
P129B	1277	4	Wolverton	1940	700322–700325	

The remaining diagrams can be divided into two sub-groups, the larger of which consisted of vehicles with square ended wells, fitted with timber baulks, supported by transverse baulks over the bogie centre line. All these vehicles had diamond framed bogies as already described, and their main dimensions are tabulated below. P126 is illustrated in Plate 25C.

Diag	Built	Length over Headstocks	Well Length	Bogie Wheelbase	Bogie Centres
P126	1925	46ft 0in	25ft 0in	5ft 6in	36ft 6in
P126B	1938	46ft 0in	25ft 0in	5ft 6in	36ft 6in
P126C	1940	48ft 0in	25ft 6in	6ft 0in	38ft 0in
P129	1926	56ft 0in	35ft 0in	5ft 6in	46ft 6in
P129B	1940	58ft 0in	35ft 6in	6ft 0in	48ft 0in

Diag	Tare with Baulks	Tare without Baulks
P126	21T 17c	20T 0c
P126B	22T 0c	20T 5c
P126C	28T 17c	
P129	24T 19c	22T 0c
P129B	36T 0c	

The remaining two diagrams had wells with curved ends, and had no provision for baulks to be fitted. Vehicles to the first of these diagrams (P126A) were built in 1930 and were 48ft 0in long over headstocks, the straight portion of the well being 25ft 6in long. The bogies were 6ft 0in wheelbase diamond framed type as before and were set at 38ft 0in centres, tare weight being 26 tons 8 cwt.

Vehicles to the second diagram (P129A) were built in 1930 and were 58ft 0in long over headstocks with a 35ft 6in well. Bogies were again of the 6ft 0in wheelbase diamond framed type set at 48ft 0in centres, and tare weight was 32 tons 8 cwt.

50 Ton Capacity

Five diagrams were issued for this capacity by the LMS. Of these, two were warwells of the type converted for use as bogie bolsters, already described. These vehicles had sloping ends to the well to enable tanks and similar vehicles to be driven directly on to them from end-loading platforms, the flat portion of the well being 12ft 0in long. The first type (diagram P133D) were used without any apparent modification, while the second (P133F) had transverse bolsters fitted for carrying locomotive boilers. The remaining three types had two six-wheeled bogies, with a wheelbase of 5ft 0in + 5ft 0in, the wheels being 3ft 0in diameter. The bogies had plate frames with screw down handbrakes and extended beyond the body frame.

Vehicles to diagram P133A were built in 1937–8 and were fitted with timber baulks. Length over headstocks was 59ft 6in, well length being 25ft 0in, and bogie centres 45ft 6in. Tare weight was 37 tons 3 cwt with, or 33 tons 10 cwt without, baulks.

Vehicles to diagram P133B were built in 1938 and were fitted with trestles. Length over headstocks was 60ft 0in with a well length of 25ft 0in and bogie centres 46ft 0in. Tare weight was 38 tons 0 cwt with, or 34 tons 13 cwt without, the trestles.

The final diagram P133C was similar to P133A and relates to wagons built in 1938. Length over headstocks was reduced to 55ft 1in, the well being 20ft 6in long with bogie centres 41ft 0in. Tare weight was 29 tons 13 cwt with, or 28 tons 11 cwt without, the baulks.

Plate 25D A 65T bogie trolley, built by Hurst Nelson in 1931; they were well wagons with detachable ends.

Summary of 50 Ton Trolley

Diag	Lot	Qty	Built	Year	Numbers
P133A	1072	1	LNER	1937	700333
	1166	4	"	1938	700337–700340
P133B	1071	3	LNER	1938	700330–700332
P133C	1130	3	Fairfield	1938	700334–700336
P133D		50			
P133F		12			360329–360340

55 Ton Capacity

Only one diagram was issued for this capacity, the vehicles to diagram P134A being built in 1938. They were very similar to the twelve-wheeled vehicles of 50 ton capacity, but the bogies had a 5ft 6in + 5ft 6in wheelbase. Length over headstocks was 68ft 0in with a well length of 32ft 0in and bogie centres of 53ft 0in. These vehicles were also provided with trestles, tare weight being 36 tons 6 cwt with, or 34 tons 11 cwt without, them.

Summary of 55 Ton Trolley

Diag	Lot	Qty	Built	Year	Numbers
P134A	1070	2	LNER	1938	700350–700351 249998

60 Ton Capacity

Here again only one diagram was issued for this capacity, the vehicles to diagram P135A being built in 1929. These vehicles were coded BTY and were used to transport large transformers and similar loads. They were mounted on four four-wheeled bogies of 5ft 6in wheelbase; these were of the diamond framed type with 2ft 8½in diameter wheels and were mounted in turn on sub-frames at 11ft 0in centres. The main 'body' consisted of a heavy girder frame 46ft 8⅛in long, mounted on the sub-frames at 40ft 0in centres, length over headstocks being 59ft 6in. Handbrakes of the bogie mounted short lever type were provided, and the vehicles tared 39 tons 19 cwt.

Summary of 60 Ton Trolley

Diag	Lot	Qty	Built	Year	Numbers
P135A	449	2	C. Roberts	1929	168906–168907

65 Ton Capacity

Two diagrams were issued for this capacity, being coded BTC. Both diagrams applied to one lot, and there appears to be some confusion as to whether two or three vehicles were involved, as either diagram could be built up using some common parts. These vehicles were well wagons with detachable ends, which consisted of six wheeled bogies of 5ft 0in + 5ft 0in wheelbase; wheel diameter was 2ft 8½in and screw down handbrakes were provided. A short heavy girder frame was mounted on top. to which the well girder portion could be attached.

These vehicles were built in 1931, the well portion of diagram P135B giving a well length of 40ft 0in, bogie centres being 58ft 0in and length over headstocks 72ft 1in; tare weight was 49 tons 8 cwt. Diagram P135C indicated a clear well length of 18ft 0in, bogie centres being 40ft 0in and length over headstocks 54ft 1in; tare weight in this case was 37 tons 7 cwt. (Plate 25D).

Summary of 65 Ton Trollies

Diag	Lot	Qty	Built	Year	Numbers
P135B	Pt593	2	Hurst Nelson	1931	5000, 5050
P135C	Pt593	1	„ „		5000

80 Ton Capacity

These vehicles were coded BTQ, and were of the well type, being mounted on four four-wheeled bogies. The bogies were of the plate framed type, with a 6ft 6in wheelbase and had short lever hand-brakes. The bogies were mounted in turn on sub-frames at 13ft 0in centres, and the heavy girder well body gave a dimension of 47ft 6in to centres of sub-frames and 73ft 8in over buffers. The well length was 22ft 0in with a flat portion of 16ft 8in. Removable timber baulks were provided, mounted on transverse baulks clear of the well.

Summary of 80 Ton Trollies

Diag	Lot	Qty	Built	Year	Numbers	Tare
P136	245	1	Hurst Nelson	1926	17000	47t 13c
	430	1	„ „	1926	327	47t 2c
P133E	1493	6	Derby	1947	700270-700275	
	1548	4	„		700276-700279	

120 Ton Trolley

This was the largest trolley built to an LMS diagram, and as can be imagined was a very large vehicle indeed. It was constructed in 1930 to diagram P136A, and had four six-wheeled bogies with channel section frames. The bogies had a wheelbase of 5ft 0in + 5ft 0in, with 2ft 8½in diameter wheels, and were coupled together in pairs by channel section girders at 15ft 0in centres. The main consisted body of two girders built up from plate, angles etc, and was 5ft 0in deep in the centre for a length of 24ft 0in; the girders then tapered on their undersides until they were about 2ft 0in deep at the extreme ends. The body girders were mounted at each end on the centre of the bogie girders at 55ft 0in centres, the inner pair of bogies thus being set at 40ft 0in centres, while the length over headstocks was 84ft 0in, and over buffers 87ft 1in. Tare weight of this vehicle was 58 tons 1 cwt, and it was thus heavier than many of the shunting engines which must have dealt with it from time to time.

Summary of 120 Ton Trolley

Diag	Lot	Qty	Built	Year	Numbers
P136A	516	1	Trade	1930	300000

Twin Girder Trucks

THESE vehicles consisted of two open frame trucks, semi-permanently coupled together, and to this end the normal couplings were dispensed with at their inner ends. Instead they had a simple link coupling, with rubbing plates which were of circular form in plan, and which served to prevent relative movement between the two trucks. The bodies were constructed from steel channels, the height from rail to the top of body being 2ft 11in; to maintain buffers and drawgear at the correct height, headstocks attached to brackets on top of the body were provided at the outer ends only.

The vehicles to diagram P137 were built in 1926 and had a carrying capacity of 50 tons (total) and were coded TGP. The trucks had four wheels 2ft 8½in diameter and 9ft 6in wheelbase. Each truck had a length from headstock to end of frame of 17ft 7¼in, and the length over headstocks for the pair was 36ft 2½in, while tare weight for the pair was 12 tons 6 cwt. The vehicles to diagram P137A were very similar in general outline; they were built in 1931 and had a carrying capacity of 70 tons (total) being coded TGT. The trucks in this case had six wheels on a 6ft 0in + 6ft 0in wheel-base. The length of each truck was 20ft 1⅛in, and length over headstocks was 41ft 3in, tare weight being 17 tons 3 cwt for the pair.

Summary of Twin Girder Trucks

Diag	Lot	Qty	Built	Year	Numbers
50T P137	298	2 pairs	Derby	1926	263369–263370 269943–269944
70TP137A	603	1pair	Derby	1931	5045–5046

Livery Details

Although there are many variations of these vehicles, their livery can be summarised as grey: large, centrally placed LMS; bauxite: small LMS to the left hand end with running number, tare, N etc., generally following the usual style, being placed to the left hand end of the vehicle. Numbers at both ends of these larger vehicles was not unknown. With bauxite livery the 'standard procedure' of smaller lettering to the left hand end was adopted although sometimes the left hand end was one third of the way along!

What is noteworthy about those built by, or for, the LNER is that, unlike most LNER vehicles, they were lettered LONDON & NORTH EASTERN RAILWAY in full along the well, and in the case of the TRESTLE gave tare weights with and without the trestles.

CHAPTER 14

CONTAINER TRUCKS AND CONTAINERS

THIS survey of standard LMS goods vehicles concludes with a description of the one item of goods equipment which, although not included in the term 'vehicles' in the strict sense of the word, has made such an impact on goods handling in this country, and which has rendered obsolete so many of the vehicle types described in this volume.

The container principle, is almost as old as steam railways themselves, for the pioneer railways of the 1830s and 1840s sometimes carried demountable containers of a primitive kind. The container, which can be carried by road or rail vehicle, did not develop at that time, and it was not until the mid 1920s that the railways, in the face of growing competition from the road haulage industry, adopted the principle in earnest. Once started, the LMS together with the other three major companies developed containers for many different types of traffic, and by the time of nationalisation in 1948 many thousands were in everyday use.

In retrospect it seems strange that the full potential of the system was not realised even then; containers were for the most part treated as ordinary vehicles, and were run in normal goods services, with all the delays arising from remarshalling and shunting which was a feature of British operation until the 1960s. Most of the photographs of the period which show block trains of containers were specially posed for the occasion, and it does not appear to have occurred to anyone that this was *the* way to run the service.

Container Trucks

IN the early stages of container development, it was anticipated that journeys would always be road-rail-road. At the same time the railways were determined that the road journey should be kept as short as possible, and many a lone container ended up at a remote branch line terminus, having waited for the daily pick-up goods to deliver it there. At many of these branch stations there would only be a horse drawn dray, while to off-load the container there would be a hand-operated yard crane, all of which with its inevitable slowness tended to negate advantages offered by the container service.

It is perhaps surprising that containers were at first loaded into ordinary open goods wagons of both five and three plank types, where the height of the sides must have added considerably to the problems of loading and unloading, although once unloaded these vehicles could of course be used for normal traffic.

Many of the vehicles used at this stage appear to have been of pre-grouping origin, most having a low carrying capacity. Some were branded 'for container traffic' although without the benefit of a diagram number, and it is not possible to be more specific about the types involved. The first mention of container trucks in the lot book was lot 346, which stated that it was for one vehicle built at Derby, no other details being available.

The first diagram was D1823. These vehicles were ex-MR three plank (medium goods) wagons, the conversion date being 1930–1. They were of the all-wood type with drop sides, 14ft 11in long over headstocks on a 9ft 0in wheelbase, being provided only with handbrakes of the double type. Carrying capacity was 8 tons, and tare weight 6 tons 16 cwt. They retained their code of MG, and it is difficult to see why they were reclassified, for little if any structural alterations appear to have been made.

These vehicles could be used for a dual role, and as the early containers could only be loaded and unloaded with ease on a road vehicle, provided an economical solution to the new system. With the introduction of more specialised containers a railhead was often available at either the beginning or end of the journey, for example at docks. Containers could thus be loaded or unloaded as a normal wagon, and for this purpose were provided with side doors, while special vehicles were developed to cater for this traffic.

The wagons were described in the diagram book as 'chassis for containers', an apt description for they consisted of a steel underframe without a floor, on the top side of which were the brackets which prevented the container from moving sideways or lengthways during transit; to hold the container down short chains with hooks and screw shackles were provided. They could not be used for any other purpose, and indeed were branded with the particular type of container they were to carry, since sizes varied and the brackets were positioned accordingly.

The first two diagrams D1838 and 1813 were for vehicles built in 1932–4; they consisted of a standard steel underframe on a 10ft 0in wheelbase. They were fully fitted, and had Morton type brakes with twin shoes on each wheel, their code being CC. The only difference between the vehicles lay in the position of the brackets, which were arranged to suit the type of container being carried. Both diagrams had examples for use with the M-type while others were used with the FX-type. Tare weight of D1838 was 5 tons 19 cwt while D1813 tared 5 tons 17 cwt. Container classifications are described later in this chapter.

The next three diagrams were for vehicles which followed the specification outlined above, but which had in addition a steel plate drop flap on each side, which could be lowered to form a bridge between container and loading dock. These diagrams were, D1975 for use with type BR containers built 1935–9, D1976 fig 35 for use with FM containers, built 1935–7 and D2065 for FM and FR type containers built 1941. All vehicles tared 5 tons 15 cwt, and were again coded CC.

The final three diagrams were very similar to those described in the previous paragraph, but the vehicles were built from second hand four-wheeled milk tank underframes. The dates when they were built are not certain but as the decision to convert all milk tanks to six wheelers was taken in 1937 it is probable that these vehicles date from that time. D2063 had standard underframes on a 10ft 0in wheelbase, and were for BM type containers. D2062 were 18ft 0in over headstocks on a 10ft 6in wheelbase and were used for class BR containers, while the final diagram D2064 was for vehicles 19ft 6in over headstocks on a 12ft 0in wheelbase for use with BM type containers. Tare weight of all these vehicles was 5 tons 15 cwt, the code again being CC.

Containers

As will be seen from the summary tables, the containers were built to a great many diagrams, and it is impossible to describe in the text the many minor variations that existed between the different batches. We shall confine our remarks here to a discussion of the main features of each type, and leave the tables to fill in the details. Both wood and steel, open and covered types were produced, one feature common to all being the provision of four lifting eyes, secured to strapping running down to floor level; this relieved the bodywork from lifting strains. Wooden containers in general followed contemporary wagon building methods,

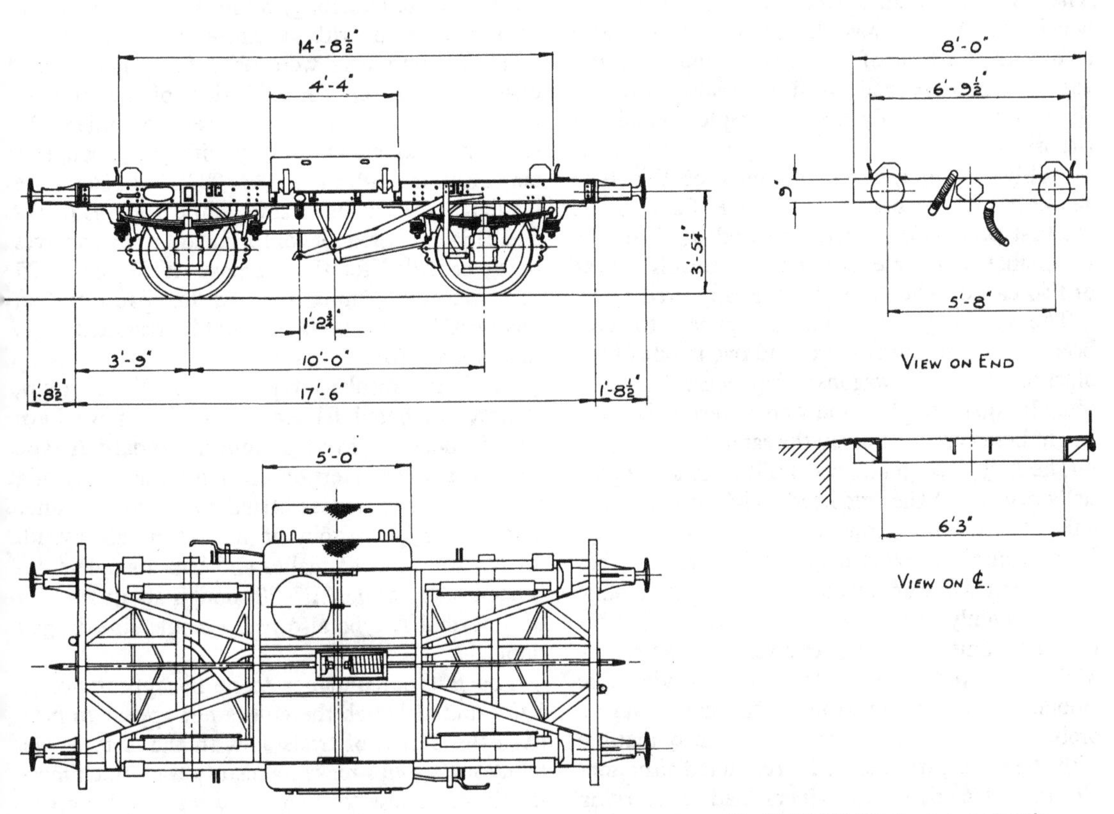

Fig 35 D1976 FM Container chassis

Summary

Diag	Lot	Qty	Built	Year	Numbers	Cont Type
1823	518	1000	Derby	1930/1		
1838	667	20	Derby	1932		M, FX
	679	20	"	1932		M, FX
1813	718	130	Wolverton	1933	} Sample 231982	M
	719	130	"	1933/4	} Sample 237061	FX
1975	890	50	Cammell Laird	1935	705070–705119	BR
	972	50	B'ham C & W	1936	705120–705169	
	1216	15	Wolverton	1939	705570–705589	
1976	888	40	B'ham C & W	1935	705000–705039	FM
	889	30	Hurst Nelson	1935	705040–705069	
	973	100	Metro Cammell	1936	705170–705269	
	974	50	Chas Roberts	1936	705270–705319	
	975	50	Hurst Nelson	1936	705320–705369	
	1057	100	Derby	1937	705370–705469	
	1058	100	Wolverton	1937	705470–705569	
2065	1301	100	Derby	1941	705590–705689	
2062	Rebuilt milk tank underframe					BR
2063	" " " "					BM
2064	" " " "					BM

the covered types having curved wooden grooves. The steel variety were constructed with steel panels, either having widely spaced corrugations and little external bracing, or with plain steel panels and heavy angle iron bracing; the roof was either flat, or a very shallow pent type.

Covered Containers

The covered containers were built in two sizes, of which the A type was the smaller. They were among the earliest of the LMS containers, and experiments were made with materials, for out of the series A1-29 only three examples remain in the diagram book. A photograph of what was probably the first LMS container of this type shows it to have been numbered A Ex1, and as the first production batch appeared in 1926 it is reasonable to assume that it was probably a year or two earlier when this example appeared.

The size of these containers appears to have been dictated by the need to load two to one of the older open goods wagons; they were therefore roughly square in plan, and were almost cubes, the height being approximately the same as the length.

The A type container was built to this size partially because of the restricted weight capacity not only of the goods crane but also road vehicles. When container traffic began to increase, further investment was worth while in heavy duty mobile cranes, mainly from Walkers of Wigan but later from Ransomes. Suitable road vehicles had meanwhile been provided by the motor trade. The mobile crane also overcame the manoeuvering problems of the larger containers, incompatible with the fixed yard crane in a restricted situation. Up to A986 these containers had a carrying capacity of 2½ tons, with the exception of A60–69, which together with those numbered from A987 upwards were rated to carry 3 tons. They had doors at one end only, in general consisting of a pair of side hinged doors for the upper part, with a drop down door below them; the exceptions were A30–69 which had full length side hinged doors only. The majority were either of wooden or steel construction; A16 and 325–331 had steel frames with Bonmax panelling, while A17 was of similar construction but with vulcanised fibre panels.

These containers were used for general merchandise, and were the equivalent of the ordinary covered goods vans; some were modified for specific workings, apparently without a change in classification: A82, 163, 188, 220, 244, 264 were insulated for banana traffic, while A91, 271, 272 and 278 were fitted for confectionary traffic, shelves being provided for this purpose. A191, 221, 225 and 252 were similarly fitted for paint traffic; finally A38 was rebuilt as a highly insulated type, renumbered AF4 in that series.

The highly insulated type was a small sub-group, at first numbered E1 etc; AF4 was, as we have already noted, converted from a standard A type container, the conversion involving the fitting of a pair of full length close fitting doors, and an inner skin giving about 6in of insulation all round. AF1–3 were much smaller and had a single side-hinged door, while AF7–28 though about the size of a normal A type also had a single side-hinged door.

The B type container made its appearance in 1927, and although there does not appear to have been the period of trials as with the A type, the many sub-groups make its history that much more involved. These containers were much larger, being more nearly equivalent to a normal covered

goods van in cubic capacity, though not in tonnage; the majority were rated at only 4 tons, probably due to yard crane capacity.

The B type as originally produced had doors at one end only, the three door type already described. This was developed into the BD type, which had additional doors of the double side hinged type in the sides. The first of these were conversions of older B type containers, B125 and 400, later examples being built new in this form. It will be noted that the B and BD types shared a

Above: Plate 26A shows the steel roll-on/roll-off container, 12 being built in 1931.

Below: Plate 26B is an example of a wooden-body furniture removal container, about 149 of this type being built between 1932 and 1933. It is mounted on an ex-LNWR dropside wagon and the original print shows the legend, 'for container traffic' on the right hand end of the vehicle. Without doubt this vehicle was modified for container traffic but never allocated an LMS diagram number.

LMS
A 118
A 134
LMS
A 134

To carry 4 Tons
DOOR TO DOOR CONTAINER
TRANSPORT
LMS
LMS
BR 15
VENTILATED
MEAT
CONTAINER
BR15
LONDON MIDLAND AND
SCOTTISH RAILWAY
N
CONTAINER ONLY
N

common number series. There were other conversions from B type, which fall into two groups. The first retained their old numbers, although some were of types which had their own series numbers, while the second group were renumbered into the correct series for the specific traffic concerned. These latter examples will be left to be described under the appropriate section, and we will deal here with the first category only. Five containers in the group B1–40 were fitted with side ventilation, the only known number being B31. Shelves were fitted to B67 and 78 for biscuit traffic, while B123 was similarly fitted for confectionary traffic. B403, 406, 547, 657, 702 and 716 were fitted with battens for furniture traffic, while finally in this group B488 was fitted with ledges for cycle traffic.

The first of the special traffic containers to be dealt with are those for the carriage of furniture. We have already noted that certain B type containers were modified for this purpose, without being renumbered. The first to carry a separate number series were also conversions; 93 containers to diagram P57 and seven to P52 were so fitted in 1929–30 and were renumbered KX1–100, the B series numbers of the P57 type are not known; the P52 conversions were B274, 275, 301, 305, 333, 345 and 352 which became KX2, 14, 16, 19, 21, 30 and 43 though not necessarily in the same order. They were of the steel type, and carrying capacity was downrated from 4 to 3 tons.

The first containers built new for this traffic were given numbers K1–300; they were of wooden constructions, and had doors of the three door type at one end only. The classification was then altered again to BK and further construction carried the numbers forward from 301; the K and BK series is thus one in the same way as the B and BD series. All K and BK containers were rated to carry 3 tons.

The next type was for the carriage of bicycles; here again one conversion from a B type has already been noted. All the other containers of this type were built new for this traffic and were designated BC; again all were of the wooden type, with end doors only. BC1–150 were of 4 tons capacity, while BC206–274 were shown as having a capacity of 1 ton 4 cwt; from this it would seem that the first group was possibly looked on as being suitable for other traffic as occasion demanded. The final group of the B type family was probably the most complex; these containers were for the carriage of meat, and apart from the several different classes considerable renumberings took place.

The first type was classed F; they were somewhat smaller than the normal B type container, and had doors at one end, and in each side, all being of the two door side-hinged type. Construction was of wood, and they were described as being insulated, although F60, 61, 93 and 234 were fitted with ice bunkers; all were rated to carry 4 tons.

Next came the BM type, at first classed M; of these 1–150 were of steel construction with wood lining; 151–175 were all-wood. All had double

Top left: Plate 26C shows two examples of wooden containers, built at Bromsgrove in 1927, a fact confirmed by the painting date on the truck. It is possible this is the prototype referred to in the chapter as being built to lot 346 but the authors are not certain.

Bottom left: Plate 26D is a good example of a ventilated container. It also shows the problem of trying to describe the various livery styles for containers owned by the Company!

Below: Fig 36 FM Container

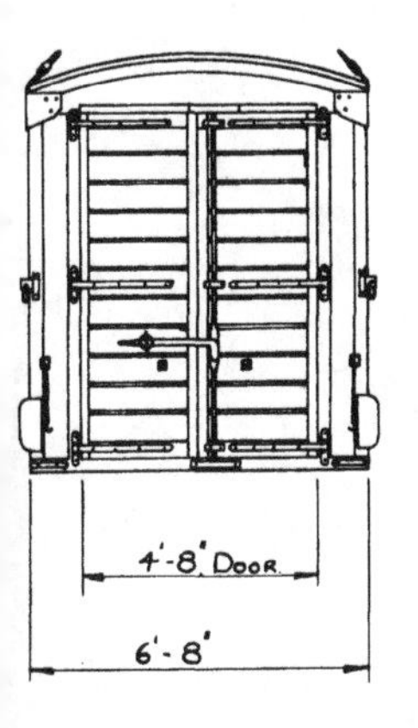

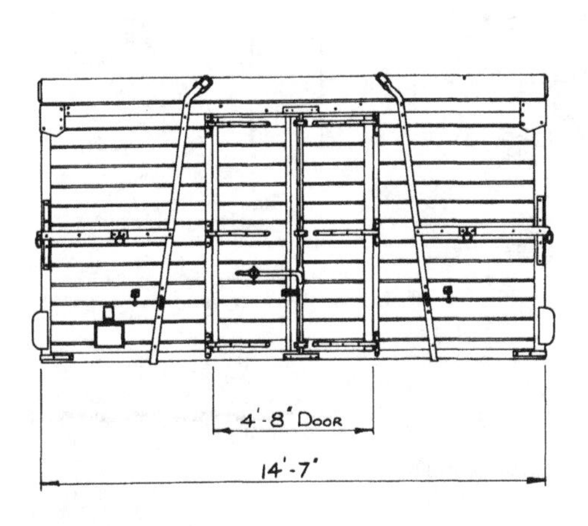

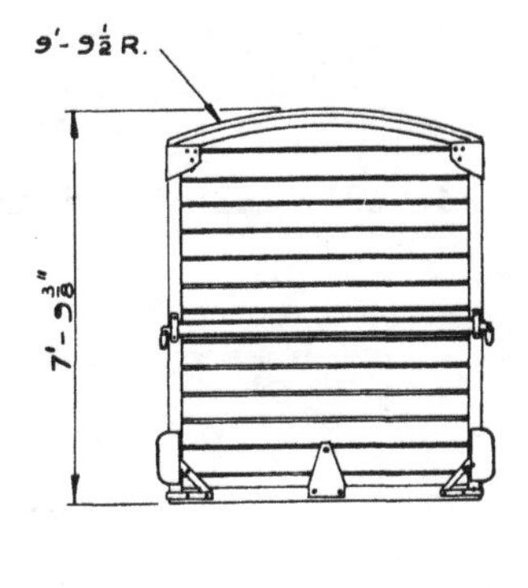

side-hinged doors in one end and both sides, and were rated to carry 4 tons. They were for fresh meat and were ventilated, ventilation taking the form of steel louvres, two to each side, while two further louvre ventilators were fitted in the doors on BM1–20, and on the body alongside the doors on the remainder.

They were followed by the BR type, which were practically identical to the BM type but had ice bunkers in the roof, BR1–100 being of plywood construction while 101–125 were of conventional timber construction; carrying capacity was again 4 tons.

The next type was the FM, (fig 36) some starting life classed FX; again they were similar to the previous types in door arrangements, but were not ventilated, and were designed to handle imported meat traffic. FM1–400 were of timber construction, 401–410 having Bonmax panels on a steel frame; they were originally FX1–10. Nos 411–420 were of the wooden type; again they were originally classed FX, carrying numbers 11–20 in that series. Nos 421–550 were also wood, and again were originally FX21–150. 551–600 were also wood originally being FX151–200; the final batch 601–956 were also of the wooden type built new as FM.

The final type in this group was the FR; they also had the same door arrangement as the other meat carriers, and were equipped with roof bunkers, being unventilated; they were described as being for imported meat or fruit traffic and were fitted with hooks for meat or movable frames for fruit traffic, carrying capacity again being 4 tons.

Open Containers

The open containers were made in three sizes, of which the H type was the smallest. This classification indicated Hod, their main use being the carriage of building materials to building sites, particularly for tall buildings where they could be lifted by crane to wherever they were required. They were of all-wood construction, no doors being fitted. Numbers H1-100 had removable lids and a carrying capacity of 1½ tons. H101–400 and 202–2064 had a carrying capacity of 2 tons and the remainder were rated at 2 tons 5 cwt.

The intermediate size of open container was the C type; again they were of the all-wood type. In this case the ends could be dropped down, while the sides were fixed into sockets, so that they could be lifted out if required; stretcher bars were fitted to assist in supporting the sides when one or both ends were in the lowered position. Carrying capacity was 2 tons for Nos 1–5, 3 tons for Nos 6–250 and 351–501 and 4 tons for Nos 251–350.

The largest of the open containers were the D (fig 37) and DX types; again they were mainly of all-wood construction, one batch only being of the all-steel type. The two classifications distinguished the door arrangements of this type. The DX containers were the same as the C type already described, with full-width drop down ends, the sides being fitted into sockets. The classification was probably applied after most of the containers were built, as photographs of the early examples show the numbers to have been D only. The D type had end doors which did not extend to the full width of the body, the sides being fixed with drop down full height doors. Carrying capacity in all cases was 4 tons.

Finally come the odd containers which do not fall into any of the foregoing categories. The first was the Bulkdrop type, of all-steel construction, which were of the open type, but the floor could be

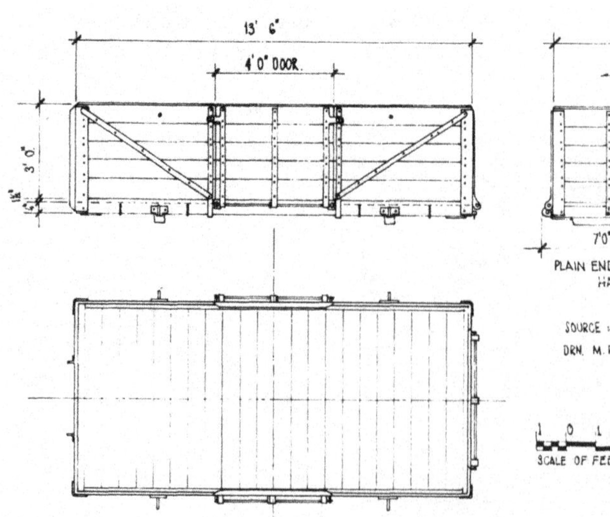

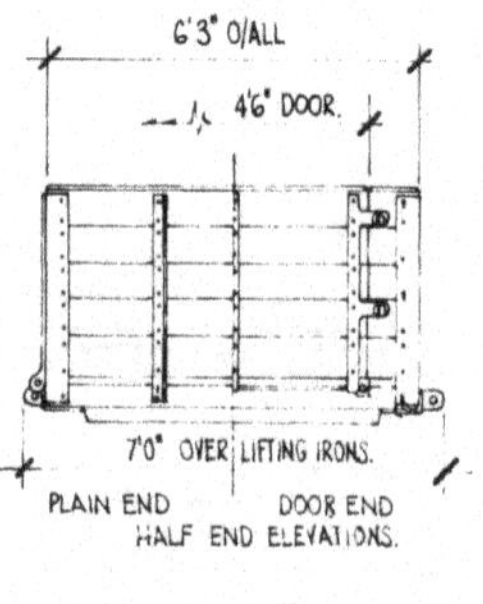

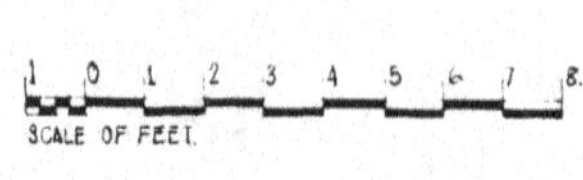

Fig 37 D Type open container

completely opened up under the control of handwheels to discharge the contents. Next come the collapsible fish containers, classed PF and used for Fleetwood-Belfast fish traffic. They were of all-wood construction, and in the normal position were 5ft 10½in long by 3ft 9in wide; overall height was 4ft 7in, reduced to 1ft 6½in when collapsed. Four small wheels were fitted on which they could be moved around.

The last of the containers have already made an appearance under the shock absorbing wagon heading; they were the steel 'roll off containers' classed ROC. The ROC container was an attempt to apply the container system to bulk minerals. The container was constructed on light weight principles employing corrugated sheeting for strength without weight. The sides were hinged at the bottom in two halves and the ends top-hinged. Thus once loaded from under a hopper the load could be discharged sideways or end ways. Even with lightweight construction the load was too heavy to be slung on a crane economically, so arrangements were made to slide it sideways from rail chassis to lorry chassis. Power for this transfer was provided by the Yorkshire steam lorry chassis with power operated two-way rams. The container was locked on the lorry chassis in such a way that it could be discharged by tipping either sideways or end ways.

Livery Details

It is an almost impossible task to cover the various livery details used by the LMS. Depending upon type and date of building crimson lake, white, grey or bauxite were used and while plate 26B is lake, 26D is white and 26A is grey. It is not certain just what colour was used for the containers shown in plate 26C. The date of the photograph is 1927 and it is possible that the containers were in the dark grey mentioned in Chapter 2 while the truck was in a lighter grey colour.

Certainly meat, ventilated, insulated type containers were white, while furniture removal containers were painted in passenger style, lake body with coach style lettering. The more general type of container was often light grey, but we have other photographs showing a different body colour which could be bauxite, or dark grey. Lettering on lake containers was yellow, on white containers black, the others being lettered in white, with black ironwork on the white body containers.

On some types the Company's name was written in full while others were merely lettered LMS.

Summary of Containers

Type A

Page	*Lot*	*Qty*	*Built*	*Year*	*Number*	*Remarks*
1		1	Butterley	1928	A13	But'ly Drawing Steel
2	HO369	1	Derby	1929	A16	Bonmax
3	HO388	1	"	1929	A17	Vulcanised Fibre
4	HO206	30	Bromsgrove	1926	A30–59	
	HO232	10	"	1927	A70–79	
5	HO220	10	"	1926	A60–69	
6	HO246	60	"	1927	A80–139	
7	HO349	50	Derby	1928	A140–189	
	HO370	75	"	1928	A190–264	
	HO453	60	Bromsgrove	1928/9	A265–324	
8	HO501	7	Derby	1929	A325–331	Bonmax
9		50	B'ham C & W	1929	A332–381	Steel Ventilated
		25	" "	1929	A462–486	
10		80	Butterley	1929	A382–461	Steel Ventilated
11		75	Metro Cammell	1929	A487–561	" "
12		105	Butterley	1930	A562–666	" "
13		120	Metro Cammell	1930	A667–786	" "
14	762	100	" "	1934	A787–886	Steel
	818	100	" "	1934	A887–986	
15	947	200	Earlestown	1936	A987–1186	
	1055	100	"	1937	A1187–1286	
16	1122	150	"	1939	A1287–1436	
17	1288	50	"	1940	A1437–1486	
	1307	159	Wolverton	1941	A1487–1645	
17	1374	200	Wolverton	1944	A1646–1757	130 LMS 70 LNER
18	1391	100	"	1945	A1776–1875	
	1395	190	"	1944	A1876–1966	91 LMS 65 LNER 31 GWR 3 SR.
	1429	300	Earlestown	1946	A1967–2266	
19	1460	100	"	1947	A2267–2366	
20	1528	150	"	1948	A2367–2516	

Type B						
50	HO244	40	Derby	1927	B1–40	
51	HO351	50	„	1928	B41–79 B92–102	
	HO371	75	„	1928	B80–91 B103–165	
	HO452	80	„	1928	B166–245	
	HO491	20	„	1929	B246–265	
52		100	Butterley	1929	B266–365	Steel Ventilated
53		50	Gloucester C&W	1929	B366–415	„ „
54		150	B'ham C&W	1929	B416–565	„ „
55		150	Metro Cammell	1929	B566–715	„ „
56		1	Butterley	1929	B716	Steel
57		351	„	1930	B1717–1067	Steel Ventilated
58	764	50	Metro Cammell	1933/4	B1068–1117	Steel
59	763	50	Metro Cammell	1934	B1118–1167	Steel
60	868	100	„ „	1935	B1168–1267	„
61	945	200	Earlestown	1936	B1268–1467	Type BD
	1056	100	„	1937/8	B1468–1567	„ „
62	1123	150	„	1938/9	B1568–1717	„ „
63	1287	150	„	1940	B1718–1876	„ „
64	1308	160	Wolverton	1941	B1868–2027	
	1375	160	LNER	1944	B2028–2187	
	1376	50	GWR	1944	B2188–2237	
	1377	50	SR	1944	B2238–2287	
65	1392	246	Wolverton	1945	B2288–2533	Type BD
	1396	267	Earlestown	1945	B2534–2800	„ „
	1397	20	GWR	1945	B2801–2820	„ „
	1398	120	Earlestown	1945	B2821–2940	„ „
	1399	36	SR	1945	B2941–2976	„ „
	1418	200	Earlestown	1946	B2977–3176	„ „
66	1461	50	„	1947	B3177–3228	„ „
67	1571	1	SLWP IOW	1948	B9999	„ „
Type BK						
100	891	100	Earlestown	1935	BK301–400	
101	1219	50	„	1939	BK401–450	
	1311	2	„	1940	BK451–452	
	1360	130	Wolverton	1943	BK453–528	
102	1393	64	„	1945	BK583–646	
Type K						
121	637	99	Earlestown	1932	K2–100	
	638	1	Derby	1932	K1	
	680	40	„	1933	K101–140	
	708	9	„	1933	K141–149	
122	722	1	„	1933	K150	
123	753	50	Earlestown	1933/4	K151–200	
	813	100	„	1934	K201–300	
Type KX						
124		93	Butterley	1929/30	KX1, 3–13, 15, 17, 18, 20, 22–29, 31–42, 44–100.	
Type BC						
125	943	50	Earlestown	1936	BC1–50	
	1053	50	„	1937	BC51–100	
126	1120	50	„	1938	BC101–150	
127	720	4	Derby	1933	BC206–209	
	752	40	„	1933	BC210–249	
	814	25	Earlestown	1934	BC250–274	
Type BM						
150	666	20	Derby	1932/3	BM1–20	
151	721	130	Pickering	1933–4	BM21–150	
152	1220	25	Earlestown	1940	BM151–175	
Type BR						
175	886	50	Earlestown	1935	BR1–50	
176	944	50	„	1936	BR51–100	
177	1121	25	„	1938	BR101–125	
Type F						
200		50	Metro C & W	1929	F1–50	
201	585	50	Earlestown	1931	F51–100	
	614	100	„	1931	F101–200	
	655	100	„	1932	F201–300	

Page	*Lot*	*Qty*	*Built*	*Year*	*Numbers*	*Remarks*
Type FM						
202	948	200	Derby	1936	FM1–200	
	1054	200	Earlestown	1937	FM201–400	
203	677	10	Derby	1932/3	FM401–410	Originally FX.
204	678	10	,,	1933	FM411–420	
205	717	130	Earlestown	1933	FM421–550	
206	885	50	,,	1935	FM551–600	
207	1300	101	D'by & W'ton	1941	FM601–701	
	1462	130	Earlestown	1947	FM702–831	
	1531	125	,,	1948	FM832–956	
Type FR						
220	884	20	Earlestown	1935	FR1–20	
Type AF						
225	HO206	1	Bromsgrove	1926	AF4	
226	1066	3	Earlestown	1937	AF1–3	
227	1431	2	,,	1946	AF7–8	
	1537	40	,,	1949	AF9–28	20 for LNER
Type C						
300	HO205	5	Bromsgrove	1926	C1–5	
301	HO247	45	,,	1927	C6–50	
302	HO492	50	,,	1929	C51–100	
	HO557	50	,,	1929	C101–150	
303	629	100	Earlestown	1931	C151–250	
304	754	100	,,	1932	C251–350	
305	816	50	,,	1935	C351–400	
	949	50	,,	1936	C401–450	
	1312	1	,,	1941	C451	
	1462	50	,,		C452–501	
Types D and DX						
325	HO201A	3	Bromsgrove	1926	DX1–3	
326	HO201A	22	,,	1926	D4–25	
	HO230	20	,,	1926	D26–45	
327	HO245	189	Derby	1927	D46–234	
328	HO350	50	,,	1928	D255–304	Type D
329	HO493	16	,,	1929	D235–250	,, ,
	HO556	4	,,	1929	D251–254	,, ,,
	HO372	50	Bromsgrove	1928	D305–354	,, ,,
	HO493	43	Derby	1929	D355–388	Type D
	HO556	91	,,	1929	D389–479	,, ,,
330	HO583	5	,,	1929	D480–484	Type D steel
	601	100	,,	1931	D485–584	
331	630	100	Earlestown	1931	D585–684	Type D
332	817	50	,,	1935	D723–772	
333	952	50	,,	1936	D773–822	
334	946	100	,,	1936	D823–922	Type D
	1481	250	,,	1947	D1031–1280	Type D
	1532	136	,,	1949	D1281–1416	
348		38			D685–722	Con from L&Y Flats
349		30			D1001–1030	
Type H						
350	602	100	Derby	1931	H1–100	
351	624	300	,,	1931	H101–400	
	PT1486	25	Bromsgrove	1947	H2020–2064	
352	706	4	Earlestown	1933	H401–404	
	707	296	,,	1933	H405–700	
	724	300	,,	1933	H701–1000	
	811	250	,,	1934	H1001–1250	
	PT1486	25	Bromsgrove	1947	H2020–2064	
Bottom Door Container						
380	1552	3	Bramilow & Edwards	1948	BULKDROP 1–3	
Type PF Collapsible Container						
400	892	1	Earlestown	1935	PF1	
	933	11	,,	1935	PF2–12	
	951	6	,,	1936	PF13–18	
Roll Off Container						
500		12		1931	ROC1–12	

Container Dimensions

Type A Steel

Numbers	Length x	Width x	Height	Tare	Cap'y cu ft.
13	7' 1⅜"	6' 7⅜"	7' 0¾"	14–1	284
332–381	7' 0⅛"	6' 6⅛"	7' 2⅜"	17–0	280
462–486					
382–461	7' 1½"	6' 7⅜"	7' 0⅝"	14–3	295
487–561	7' 0⅛"	6' 6½"	7' 3⅛"	13–0	298
562–666	7' 1⅛"	6' 7⅛"	7' 1¼"	13–3	300
667–786	7' 0⅛"	6' 6⅛"	7' 3⅛"	15–2	296
787–986	7' 0⅜"	6' 6½"	7' 7½"	14–3	285

Type A Wood,

Numbers	Length x	Width x	Height	Tare	Cap'y cu ft.
30–59	7' 1½"	6' 1½"	6' 11⅝"	11–2	269
70–79				13–2	
60–69	7' 1½"	6' 1½"	7' 0⅜"	13–1	269
80–139	7' 2⅜"	6' 1¾"	7' 2⅛"	16–0	263
140–324	7' 2¾"	6' 7¾"	7' 2⅜"	17–1	284
987–1286				19–2	329
1287–1436				19–2	322
1437–1775	7' 6"	7' 0"	7' 11½"	1–0–0	318
1776–2266				1–1–3	322
2267–2366				1–1–2	322
2367–2516				1–2–0	329

Type A Steel Frame with Bonmax Panels,

Numbers	Length x	Width x	Height	Tare	Cap'y cu ft.
16	7' 1¾"	6' 6¾"	7' 2⅜"	15–2	267
325–331					292

Type A Steel Frame with Vulcanised Fibre Panels,

Numbers	Length x	Width x	Height	Tare	Cap'y cu ft.
17	7' 1¾"	6' 6¾"	7' 2¾"	15–1	264

Type B Steel,

Numbers	Length x	Width x	Height	Tare	Cap'y cu ft.
266–365	13' 11⅛"	6' 7⅞"	7' 2¾"	1–3–0	577*
366–415	13' 9⅞"	6' 5⅜"	7' 4¼"	1–3–0	588
416–565	13' 9⅞"	6' 6⅛"	7' 4½"	–4–3	595
566–715	13' 10¼"	6' 7¾"	7' 3½"	19–1	589
716	13' 11⅛"	6' 7¾"	7' 3¼"	1–2–0	582
717–1067	13' 11"	6' 7⅞"	7' 2¾"	1–0–1	587*
1068–1117	13' 10¼"	6' 6¼"	7' 5"	1–4–0	592
1118–1167				1–2–3	
1168–1267		6' 6½"		1–4–0	

*Conversions to KX type Tare 1–4–0

Type B & BD Wood.

Numbers	Length x	Width x	Height	Tare	Cap'y cu ft.
1–40	13' 6"	6' 1¾"	7' 4⅛"	1–3–3	510
41–265	14' 0"	6' 7¾"	7' 4½"	1–7–2	574
1268–1567					723*
			7' 11⅛"	1–13–0	709*
1568–1717					703*
1718–1867			7' 11½"	1–14–3	703*
1868–2287				1–14–0	703
2288–2533	16' 0"	7' 0"	7' 11⅝"	1–16–3	717*
2534–2800				1–17–1	171*
2801–2820				1–15–0	717*
2821–2940				1–17–1	717*
2941–2976				1–18–0	717*
2977–3176				1–18–0	717*
3177–3226				1–18–1	717*
9999				1–13–0	743*

*Type BD.

Type K & BK Wood,

Numbers	Length x	Width x	Height	Tare	Cap'y cu ft.
1–149	15' 2¼"	6' 10½"	7' 10½"	1–9–2	650
150	15' 2"	6' 10"	7' 9⅞"	1–8–3	650
151–300	15' 2¼"	6' 10½"	7' 7⅜"	1–12–0	650
301–400			7' 9¾"	1–10–1	718*
401–582	16' 0"	7' 0"	7' 11½"	1–9–0	713*
583–646			7' 11⅝"	1–12–0	708*

*Type BK.

Type BC Wood

Numbers	Length x	Width x	Height	Tare	Cap'y cu ft.
1–100	16' 0"	7' 0"	8' 4"	1–9–3	740
101–150					
206–274	14' 4¾"	6' 10¾"	8' 3¾"	1–11–2	725

Type F Wood

Numbers	Length x	Width x	Height	Tare	Cap'y cu ft.
1–50	12' 7"	6' 8"	7' 4"	1–15–2	438
51–300			7' 6¼"	1–18–3	

Type BM

Numbers	Length x	Width x	Height	Tare	Cap'y cu ft.
1–20*	15' 7¾"	6' 9½"	8' 5¾"	2–9–0	
21–150*				2–0–0	
151–175	16' 0"	6' 11½"	8' 4⅜"	2–3–3	760

*Steel, remainder wood.

Type BR Wood

Numbers	Length x	Width x	Height	Tare	Cap'y cu ft.
1–50			8' 7⅛"	2–12–0	
51–100	16' 0"	6' 11½"	8' 10⅛"	2–10–3	760
101–125			8' 10⅛"	2–12–2	

Type FM Wood

Numbers	Length x	Width x	Height	Tare	Cap'y cu ft.
1–400	14' 7"	6' 8"	7' 9⅜"	1–17–0	597
401–410*	14' 1⅞"	6' 2⅞"	7' 7½"	1–18–0	542
411–420			7' 9⅞"	2–4–0	582
421–550			7' 9⅜"	1–16–0	597
551–600	14' 7"	6' 8"	7' 9⅜"	1–19–2	597
601–701			7' 9½"	2–1–0	600
702–956			7' 9½"	2–5–0	600

*Bonmax panels.

Type FR Wood

Numbers	Length x	Width x	Height	Tare	Cap'y cu ft.
1–20	14' 7"	6' 8"	7' 9¾"	2–4–3	567

Open Containers

Type C Wood,

Numbers	Length x	Width x	Height	Tare	Cap'y cu ft.
1–5			3' 5½"	8–1	126
6–50			4' 4⅝"	9–2	157
51–150	7' 1¾"	6' 1¾"	3' 8"	9–2	126
151–250			3' 8¼"	12–0	124
251–350	7' 6½"	6' 1¾"	3' 9¼"	12–0	126
351–450				12–0	
451	7' 6"	6' 3"	3' 6"	11–2	125
452–501				14–0	

Type D & DX Wood,

Numbers	Length x	Width x	Height	Tare	Cap'y cu ft.
1–3	12' 3¼"	6' 2¼"	3' 9⅞"	13–0	215
4–45	12' 4¼"	6' 5¼"	3' 11⅞"	15–0	245
46–234	13' 6"	6' 3"	4' 5½"	17–0	300
235–254	13' 6"	6' 3"	3' 8⅜"	19–2	240
255–304	13' 6"	6' 3"	3' 9¼"	1–3–0	238
305–479	13' 6"	6' 3"	3' 8⅜"	19–2	240
480–584*	13' 6"	6' 0⅛"	3' 9⅞"	18–3	238
585–684	13' 6"	6' 3"	3' 8⅜"	1–0–3	240
685–722	12' 4"	6' 3"	3' 9¼"	12–3	230
723–772	14' 0"	6' 3"	4' 0⅞"	19–2	296
773–822	14' 0"	6' 3"	4' 1¼"	19–3	298
823–922				1–0–2	
1031–1280	14' 0"	6' 3"	4' 1¼"	1–6–2	298
1281–1416				1–3–0	
1001–1030	12' 4"	6' 3"	3' 10"	14–2	228

*Steel

Type H Wood,

Numbers	Length x	Width x	Height	Tare	Cap'y cu ft.
1–100	7' 4"	4' 4"	2' 0"	6–2	35
101–400	7' 0"	4' 0"	2' 5¼"	5–2	39
401–2000	7' 0"	4' 0"	2' 4⅛"	3–3	42
2020–2064	7' 0"	4' 0"	2' 5¼"	3–3	39

APPENDIX 1

FREIGHT STOCK COLOUR SHADE

The basic composition of colour of LMS freight stock is detailed in the Livery section within Chapter One but this does not tell the reader the exact shade employed. Fortunately L. Tavender of the Historical Model Railway Society was able to examine paint samples taken from Ex LMS Mineral Wagon DM 40503 when in use as a crane runner and below are his findings—earliest colour first.

		Coats	
Assumed LMS Grey	Bluish Grey 10B 5.5/1	1	Medium Sea Grey No 637 of BS318C 1964
	Black	1	
	Moderate Reddish Brown 10R 3/4	1	Red oxide No 446 of BS381C 1964 (near)
Assumed LMS Bauxite	Strong Reddish Brown 10R 3/8	2 thick	Venetian Red, No 445 of BS381C 1964
Surface Appearance	Moderate Reddish Brown 2.5YR 3.5/4		Camouflage Red No 435 of BS381C 1964

Plate 27 A southbound freight heads through Arnside in the 1930s The leading wagons include various types of bogie bolsters loaded with rails with four-wheeled low sided wagons between as under-runners. L&GRP

APPENDIX 2

LMS BUILDING PROGRAMME 1924–47

Year	Goods Brakes	Open Goods	Covered Goods	Mineral	Sand	Tube	Long Low Plate	Ballast	Sleeper	Loco Coal	Single Bolster	Double Bolster	Deal	Banana	Gun-powder	Venti-lated Meat
1924	181	12 000	600	4 500											10	
1925	458	12 400	2 022	3 350		250						162		200	40	
1926	331	5 100	1 133	4 150								38		350		
1927	300	10 000	1 690	2 500			187					501		250		300
1928	102	7 000	1 375	4 500								205	150			
1929	200	7 650	2 250									206		100	25	
1930	225	5 750	2 700	2 000			400					500		100		150
1931	175		1 750													50
1932	9		1 755													
1933	200	100	1 550			100	250						50		20	
1934	50	3 900	3 000		100	100										
1935	90	1 700	4 000	2 500		100										
1936	120	4 750	2 500	2 500		500	525	60	20	550	500	600			20	
1937	160	6 637	250	2 000			250	500	14	440		50	50		25	
1938	300	2 638	1 000	1 000		150		1 000	75		500	150				
1939	1	2 600	750			150	200	175	50		250				20	
1940	394	300	2 435	550			100	175		254	162					
1941	128		715					600	100							
1942	225	250	3 990			100	300	250				200				
1943	225	1 750	3 256												20	
1944	250	775	2 040	1 500			400	100	100						20	
1945	191	2 300	1 550	2 000						150		200				
1946	150	1 750	500	2 000			100	200	50	150		150		100		
1947	46	2 700		2 600		300		100	50	50		50				
Total	4 511	92 100	42 991	37 650	100	1 750	2 712	3 160	459	1 594	1 412	3 012	250	1 100	200	500

Year	Refrig	Beer	Cattle	Tanks	Hoppers	Bogie Bolsters	Armour Plate Trucks	Bogie Trollies	Twin Girder Trucks	Imple-ment Trucks	Trac-tion Trucks	Deep Case Trucks	Glass Trucks	Four Wheel Trollies
1924	100		670		200							3		
1925	50		908		50			8		11				6
1926	250		433	2		89	12	10	2 pr				2	
1927	50		784	3		100							38	
1928			750	5	45	88		3		15		4		3
1929	200	100	295	9	430	12		8		5	40			9
1930	100		545	12	100			17						6
1931			150					3	1 pr	10	20			
1932			75		108									
1933			525		10									
1934			50		365									
1935			100		10	20					20			10
1936					910	83					10			
1937					175	2	7	17		10	25		10	
1938				1	500	53	1	21						
1939				6		55	10	12			10			4
1940				4	10	70		20		10			10	
1941						25		50						
1942						25		2						
1943														
1944				1						30				
1945														
1946														
1947				2		50							6	
Total	750	100	5 285	45	2 913	672	30	171	3 pr	91	125	7	66	38

Left: Plate 28 A general freight on the Midland Division in the 1920s with a mixed load of covered vans and open wagons of various types most sheeted over. L&GRP

Above: Plate 29 Toton hump yard in the 1930s with a coal train being positioned for marshalling by a then new diesel shunter. L&GRP

Year	*Chaired Sleeper Trolley*	*Match Wagons Cranes*	*Container Trucks*	*Annual Total*
1924				18 264
1925				19 915
1926				12 082
1927				16 703
1928	45			14 290
1929				11 539
1930			500	13 105
1931		6	500	2 665
1932			40	1 987
1933			130	2 935
1934			130	7 695
1935			120	8 670
1936			250	13 898
1937		1	200	10 823
1938				7 439
1939			15	4 308
1940				4 494
1941			100	1 718
1942		11		5 353
1943				5 251
1944				5 216
1945				6 391
1946				5 150
1947				5 954
Total	45	18	1 985	205 845

APPENDIX 3

BR BUILT VEHICLES TO LMS DESIGN

This appendix is not claimed to be exhaustive. It covers only those vehicles which are known to be LMS designs and where examples were built by the LMS. Thus the 12 ton BR Pipe Wagon—diagram 460—is not included because although an LMS design, the first example (to BR lot 2004) did not appear until 1949. Some common designs are also omitted. Certain vans for instance were built by the Southern Railway for itself, and for the LMS, and also appeared under BR auspices; some special wagon designs were common to the LMS and LNER and also appeared under BR. For these and other reasons, the list is not exhaustive.

BR Lot	*Qty*	*Description*	*BR diag*	*LMS*	*Diagram*	*Built*	*BR Numbers*
2001	1300	Goods Van 12 ton VB	200	95	D2108	Wolverton	750000–1299
2	80	Banana Van 10 ton VB	240	5A	D2111	"	880000–079
8	7	Flat Wagons 40 ton ARM ME	SP001	2B	Special	Derby	907100–106
9	40	Bulk Grain Van 20 ton	270	28	D1689	"	885000–039
10	50	Bogie Plate 42 ton	490	18B	Special	"	947000–049
12	150	Ballast Wagons 12 ton SOLE	565	55A	D2095	Wolverton	DB982000–149
15	200	Banana Van 10 ton VB	240	5A	D2111	"	880080–279
16	7	Flat Wagons 40 ton ARM ME	SP001	2B	Special	Derby	907107–113
18	200	Fruit Vans 12 ton VB	230	9R	D2112	"	875000–199
25	125	Goods Brake 20 ton AVP & gauge (*29/10/52 revised to diagram 505)	503*	2D	D2068	"	950000–124
26	125	Goods Brake 20 ton Unfitted	503		D2036	Derby	950125–249
28	2	Ballast Plough Brake	596		D2025	"	DB993700–01
30	10	Flat trollies 4 wheel 20 ton	SP512	96A	Special	"	900000–09
2037	250	Plate Wagons 4 wheel 22 ton	430		D2083	Shildon	930000–249
2104	1500	All steel Mineral Wagons	106		D2134	Derby	64000–65499
2108	397	Medium Goods 13 ton unfitted	017		D2101	Wolverton	457200–596
2117	11	Hoppered Ballast unfitted 25 ton	580		D1800	Metro-Cammell	DB992024–034
2118	11	Hoppered Ballast unfitted 25 ton	580		D1800	" "	992035–056
2132	300	Plate 4 wheel 22 ton	430		D2083	Sheldon	830250–930549
2135	250	Fruit vans ventilated 12 ton	230		D2112	Faverdale	875300–549
2151	500	Plate Wagons 4 wheel 22 ton	430		D2083	G. R. Turner	930550–931049
52	25	Medium Goods S/A 13 tons VB	018		D2152	Shildon	474800–474824
2172	7	Ballast Plough Brakes 16 ton	596		D2025	R. M. Pickering	DB993702–708
2174	9	Bogie Armed Plate 55 ton (ARMET, ARMD)	SP002	3A	Special	Teeside B & E	900800–808
2184	500	Coal—not hoppered All Steel (originally 2,000)	106		D2134	Derby	67600–68099
2185	50	Bogie Plate (unfitted) E 42 tons	490	18B	Special	Derby	947125–174
2187	38	Machine Low 25 ton VB	SP242	54A	Special	P. W. McLellan	904500–537
2188	30	Hopper Ballast 25 tons	580		D1800	Metro-Cammell	DB992117–196
2210	800	Coal—not hoppered All Steel (originally 1,500)	106		D2134	Derby	68100–68899
2215	50	Sleeper Wagon 12 ton	621		D1953	Butterley	DB995000–049
2264	200	Ballast Wagons 12 ton SOLE	565		D2095	Fairfields S & E	DB982150–349
2344	20	Bogie Plate (unfitted) E 42 tons	490	18B	Special	Sheldon	947175–194
2354	5	Trolley Flat 20 ton unfitted FLATROL M/V	SP512	96A	Special	Lancing	900010–014
2356	10	Trolley well (Wellrot MA) 20T	SP733	96B	Special	"	900700–900709
2450	5	Trolley Flat 20 ton unfitted FLATROL M/V	SP512	96A	Special	Derby	900015–019
2455	40	Bogie Plate Boplate E unfitted 42 ton	490	18B	Special	"	947195–234
2456	2	Bogie Plate Boplate E (for S & T)	490	18B	Special	"	947195–234
2456	2	Bogie Plate Boplate E (for S & T)	490	18B	Special	"	DB997400–401
2491	6	Glass Wagon 12 ton unfitted (GLASS MD)	SP171	75	Special	Swindon	902006–011
2543	13	Glass Wagon 12 ton unfitted & EP	SP171	75	Special	"	902012–024
2613	3	Flatrol MVV 20 ton	SP512	96A	Special	Derby	900020–022
2622	7	Armour Plate 55 ton (ARM ET)	SP002	3A	Special	Teeside B & E	900809–015
2641	4	Flatrol MVV 20T 20T	SP512	96A	Special	Derby	900023–026
2643	10	Armour Plate (ARM EL) 40T	SP001	2B	Special	Cambrian	908500–509
2646	10	Glass MD & EP 12T	SP171	75	Special	Swindon	902025–034
2651	4	Armour Plate (ARMET) 55 tons	SP002	3A	Special	Teeside B & E	908016–019
2721	2	Armour Plate (ARM EL) 40T	SP001	2B	Special	Cambrian	908510–511
2722	40	Bogie Plate (Boplate E) 42T	490	18B	Special	"	947310–349
2874	3	Glass WE 12T	SP171	75	Special	Swindon	902035–037
2924	1	Armour Plate (ARM WE) 40T	SP001	2B	Special	Cambrian	908512
2930	50	Bogie Plate (BOPLATE E) 42T	490	18B	Special	"	947350–399
2938	1	Armour Plate (ARM WF) 55T	SP002	3A	Special	Teeside B & E	908020
2945	6	Flatrol (FLATROL MVV) 20T	SP512	96A	Special	Lancing	900037–042
3041	1	Armour Plate (ARM WE) 4T	SP001	2B	Special	Cambrian	908513
3046	160	Bogie Plate (BOPLATE E) 42T	490	18B	Special	"	947400–559
3055	150	Bogie Plate (BOPLATE E) 42T	490	18B	Special	Hurst Nelson	947560–709
3056	150	Bogie Plate (BOPLATE E) 42T	490	18B	Special	R. Y. Pickerry	947710–859
3057	1	Armour Plate (WF) 55T	SP002	3A	Special	Teeside B & E	908021
3235	125	Bogie Plate (BOPLATE E) 42T	490	18B	Special	Derby	948010–134
3236	125	Bogie Plate (BOPLATE E) 42T	490	18B	Special	"	948135–259
3271	6	Armour Plate (ARM AB) 40T	SP001	2B	Special	Standard Wagon	098514–519

INDEX

CPSIA information can be obtained
at www.ICGtesting.com
Printed in the USA
LVHW060931190523
747334LV00014B/40

9 780715 373576